Algebra Refresher Workbook

for Adults Returning to School

Table of Contents

Why This Workbook is the Best Way to Learn Algebra

The Importance of Algebra in Real Life

The word "Algebra" brings back memories of fear and confusion to many adults. Equations, formulas and confusing variables seemed like a daily struggle to most, while a few others that grasped the concepts could not understand why anyone needs this stuff in daily life. However, the truth is that Algebra is not just a subject for academia – it is a practical tool used in daily life, from financial planning to problem solving at work.

Algebra is a branch of mathematics that involves use of symbols and letters to find patterns and relationship between numbers. It helps develop logical thinking, problem-solving skills, and the ability to analyze complex situations. Algebra plays a critical role in helping to make informed decisions, whether it's to calculate interest on a loan or deciding on a daily budget.

Where Algebra Is Used in Real Life

You may not realize it, but algebra is used all around us. Here are some common areas where algebra is essential:

1. **Personal Finance:** From managing credit card debt to understanding mortgage rates, algebra helps individuals calculate interest rates, loan payments, and investment growth.

2. **Business and Careers:** Algebra is a fundamental skill in careers such as engineering, economics, computer programming, and even marketing. Business owners use algebra to forecast profits, manage expenses, and analyze data.

3. **Technology and Science:** Many advancements in medicine, physics, and engineering rely on algebraic formulas to model real-world phenomena.

4. **Daily Problem-Solving:** Whether it's determining the best cell phone plan, comparing grocery store prices, or figuring out how much paint is needed for a room, algebra helps in making logical decisions.

5. **Home Projects and DIY Tasks:** Algebra is used in home renovations, sewing, carpentry, and other practical applications where measurements and calculations are needed.

Despite its usefulness, many adults find themselves struggling with algebra when they need to use it again. This leads us to the challenges of re-learning algebra later in life.

The Challenges of Re-Learning Algebra as an Adult

Returning to algebra as an adult can be intimidating. Unlike high school students who are in a structured learning environment, adults often have to juggle work, family, and other responsibilities. Here are some key challenges:

1. **Memory Gaps:** Many adults haven't used algebra for years, making it difficult to recall concepts they once learned.
2. **Lack of Time:** Unlike students who have dedicated study hours, adults often struggle to find time for learning.
3. **Math Anxiety:** Some people have negative experiences with math from their school years, leading to a lack of confidence.
4. **Poorly Designed Learning Materials:** Many resources are designed for students rather than adults, making them frustrating and difficult to follow.

This workbook is designed to overcome these challenges by providing clear explanations, real-world examples, and practical exercises tailored for adult learners.

What's Wrong with Current Algebra Textbooks?

Many traditional algebra textbooks fail to meet the needs of adult learners. Here's why:

1. **Too Methodical and Formulaic:** Many textbooks present algebra in a rigid, abstract way with a focus on memorizing formulas rather than understanding their application. Without real-world applications, these books make algebra feel disconnected from everyday life.

2. **Lack of Examples:** Some textbooks list formulas without providing step-by-step examples, leaving learners confused about how to apply them. In cases where they do, there is no sufficient explanation, which leads to confusion.

3. **No Progress Tracking:** Traditional textbooks rarely offer ways to gauge progress, making it difficult for learners to know if they are improving.

4. **No Space to Work on Problems:** Many textbooks are filled with dense explanations but lack space for practicing problems, forcing learners to use separate notebooks and making the process more cumbersome.

What's Wrong with Other Algebra Workbooks?

We have recently noticed that there are several Math and Algebra workbooks on Amazon that aim to solve the above problems. They attempt to explain Algebra in a step-by-step workbook like manner. While these Algebra workbooks aim to provide a more hands-on approach, many still have significant flaws. Here's what you may have encountered in other workbooks:

1. **Hard-to-Follow Examples:** Some workbooks provide examples, but they are overly complex or assume prior knowledge that learners may not have. They do not provide explanations for each step.

2. **Insufficient Space for Work:** A workbook should provide ample space for solving problems. Unfortunately, many only offer a few lines, requiring learners to use extra paper, which defeats the purpose of a workbook.

3. **Incorrect Answers and No Explanations:** Many workbooks provide answer keys, but some of the answers are incorrect or lack detailed explanations, making it difficult to understand mistakes.

This workbook is written by a Math tutor with over 10 years' experience tutoring students of all ages. The main goal of this workbook is to make you comfortable with Algebra. It is written in a step-by-step manner with progress goals, and plenty of examples.

The 30-Day Algebra Challenge

One of the biggest challenges adults face when re-learning algebra is not knowing where to start. Without a clear learning goal, it's easy to feel lost or overwhelmed. This book is designed to help you build a strong foundation in math by following a 30-day structured learning plan.

Each chapter is filled with practical examples and real-world applications to make learning more engaging. At the end of each chapter—and again at the end of the book—you'll find worksheets to test your understanding. These exercises are designed to reinforce what you've learned and help you track your progress.

You can choose how to approach the exercises:

1. If you're confident in a topic, you can skip directly to the algebra worksheets to test yourself.
2. If you're struggling with a concept, go through the relevant chapter thoroughly before attempting the exercises.

By the end of the book, you'll have a clear idea of your strengths and weaknesses, allowing you to focus on areas that need improvement. The immediate feedback provided by the exercises give you some extra confidence in your skills.

This book is meant to reinforce the fundamentals of Algebra, and build a stable foundation to work on. It's suitable for adults who want to rebuild their math skills, as well as students of any age who need extra support.

Even students studying advanced topics like Calculus or Differential Equations have found this book useful. Many realized they were missing key Algebra skills and used this book to quickly review and strengthen their understanding, allowing them to catch up in their courses.

For most learners, completing the full 30-day challenge has been a game-changer. By working through each concept and applying what they've learned, they have gained the confidence and skills needed to succeed in their Algebra journey—whether for personal growth, further education, or career advancement.

Before we get started, let's plan out our next 30 days. We have 30 worksheets that are designed for the next 30 days.

So, you're going to fill out the below chart as you learn the basics and complete the worksheets and exercises. Fill in the date when you complete each exercise and keep track of your progress.

Count	Date	Topic	Worksheet	Status
Day 1		Variables	Worksheet 1	
Day 2		Word Problems	Worksheet 2	
Day 3		Solving Basic Equations	Worksheet 3	
Day 4		Checking Equations	Worksheet 4	
Day 5		Laws of Algebra	Worksheet 5	
Day 6		Powers and Indexes	Worksheet 6	
Day 7		Types of Equations	Worksheet 7	
Day 8		Graphing Points	Worksheet 8	
Day 9		Points on a Line	Worksheet 9	
Day 10		Slope of a Line	Worksheet 10	

Count	Date	Topic	Worksheet	Status
Day 11		y-Intercept of A Line	Worksheet 11	
Day 12		Plotting Lines on A Graph	Worksheet 12	
Day 13		Solving Basic Equations	Worksheet 13	
Day 14		Equation of a Line	Worksheet 14	
Day 15		Fractional Equations 1	Worksheet 15	
Day 16		Fractional Equations 2	Worksheet 16	
Day 17		Simultaneous Equation 1	Worksheet 17	
Day 18		Simultaneous Equation 2	Worksheet 18	
Day 19		Group Factorization	Worksheet 19	
Day 20		Quadratic Factorization	Worksheet 20	

Count	Date	Topic	Worksheet	Status
Day 21		Factorization – Other	Worksheet 21	
Day 22		Quadratic Equations - Factorization	Worksheet 22	
Day 23		Quadratic Formula	Worksheet 23	
Day 24		Graphing Quadratics	Worksheet 24	
Day 25		Graphing Circles	Worksheet 25	
Day 26		Algebraic Inequalities	Worksheet 26	
Day 27		Quadratic Inequalities	Worksheet 27	
Day 28		Exponential Equations	Worksheet 28	
Day 29		Logarithmic Equations	Worksheet 29	
Day 30		Capstone Worksheet	Worksheet 30	

The Need for Algebraic Equations

What is An Equation?

Before we start on Algebra, let's get down to the very basics. What is an Equation in Mathematics?

An equation is basically two Mathematical statements that are equal to each other. It typically includes numbers and operations like addition or multiplication. For example:

$5 + 2 = 7$

The above equation has two statements.

The first statement is $5 + 2$. This is called the Left-Hand Side of the equation (or LHS)

The second statement is 7. This is called the Right-Hand Side of the Equation (or RHS).

In an equation, the LHS and RHS are always equal to each other.

So, $5 + 2 = 7$.

What is an Algebraic Equation?

When an equation includes letters in either the LHS or RHS it becomes an Algebraic Equation.

This helps solve real-world problems where values are missing or changing.

A simple algebraic equation is below:

x + 5 = 10

x + 5 is the LHS and 10 is the RHS.

Parts of an Algebraic Equation

In the equation x + 5 = 10, the different parts are:

- **Variable (x):** Represents an unknown number
- **Constant (5 and 10):** Fixed numbers in the equation
- **Operator (+):** Shows the mathematical operation
- **Equals sign (=):** Indicates both sides are equal

Solving An Algebraic Equation

An algebraic equation can be solved when one important condition is met.

No. of variables = No. of equations

In equation, x + 5 = 10, there is:

1 variable (x)

1 equation

So, this equation can be solved since number of variables is equal to the number of equations.

Now, let's look at an equation that cannot be solved.

$x + y = 5$

There are 2 variables (x, y) and 1 equation.

No. of variables $\neq$ No. of equations

So, this equation cannot be solved. It is called a multivariable equation. To solve this, we need additional information. We will cover solving these types of equations in future chapters.

So, let's start off our first worksheet to check if a set of equations can be solved are not.

The answers for all worksheets are at the back of the book.

Worksheet 1

Can the below equations or sets of equations be solved?

1. $x + 2 = 3$

$$x = 1 \quad Y.$$

2. $x + y = 5$

NO

3. $2x + y = 3$
   $x + 2y = 4$

$2 + 1$

$x = 1 \quad y = 1.$ NO

$1 + 2$

1×2

$x = 2$

4. $x + y + z = 3$
   $x + 2y + z = 5$

no ~~counke~~ 1 or 0

5. $x + y + 2z = 3$
   $x + 3y + z = 5$
   $x + 7y + 2z = 4$

Now, once we know which equations can be solved, how do we solve an Algebraic Equation?

Let's look at below equation:

16

$x + 5 = 10$

First, we need to figure out what needs to be solved. In above equation, the value of the variable x is unknown. So, we solve this equation when we find the value of x.

Next, we need to get all the constant terms to one side. In above equation, we have two constants 5 and 10. Let's get them both on the RHS.

$x + 5 = 10$

To do so, we can subtract 5 from both LHS and RHS of the equation. When we subtract any term from both sides of the equation, the equation remains the same.

$x + 5 - 5 = 10 - 5$

$x = 5$

So, $x = 5$ is the solution to our equation.

How Does This Equation Apply to Daily Life?

Now, if you're new to Algebra, the equation $x + 5 = 10$ might seem pointless. Why do we even need this equation? Why do we need to find the value of x.

The reality is, the above algebraic equation is derived from a real-life example, and is used in everyday problem-solving, budgeting, and planning. In above example, if you have $10 and spend $5 on snacks, how much do you have left? This is represented as:

$5 + $(Money Remaining) = $10

Now, we represent money remaining as $x, as it's easier to work with one letter instead of an entire phrase. So, the equation becomes,

x + 5 = 10

Solving for x as done previously shows that x = $5. So, you have x = $5 remaining.

As you can see, it helps to write out a problem statement as an equation so it's easier to solve. It makes even more sense when you have larger more complicated equations with multiple variables as you will see later in this book.

But just to hammer this point home, let's look at a few more examples.

Shopping Discounts

If a shirt costs p dollars and there is a $10 discount, the final price F is:

$F = p - 10$

If the original price p was $50, then:

F = 50 – 10 = 40

F = $40

Speed and Distance Calculation

If a car travels at 60 km/h and you want to know how long it takes to travel 120 km? We can solve this using the Algebraic Equation:

$$T = \frac{D}{S}$$

Where T is the time taken, D is the distance travelled and S is the speed.

$$T = \frac{120}{60} = 2 \ hours$$

The time taken is 2 hours.

Saving Money

If you save $20 per week for w weeks, the total savings S is given by the equation:

$$S = 20w$$

Where S is the total savings, and w is the number of weeks.

If you save for 10 weeks:

$$S = 20 \ x \ 10$$

S = 200

So, the savings is $200 after 10 weeks.

Algebra helps us solve real-life problems efficiently!

Figuring Out Variables.

So, let's look at all the above equations again:

x + 5 = 10

F = p – 10

$$S = 20w$$

What do s, F, p, w and x have in common? They are all the unknown values in the equation. We represent them with letters or symbols, known as variables.

Now, let's go ahead and do this with another problem:

Seven more than a number is 13. In Algebra, the unknown number is the variable. We can represent a variable with any letter or symbol. Let's say we choose the letter 'x'. Then, the Algebraic equation becomes:

x + 7 = 13

Now, in another example, we are told that 4 times a number divided by 3 is 10. So, let's choose another letter as the variable. Let's say "a" is the variable. Then, the Algebraic Equation becomes:

$$\frac{4a}{3} = 10$$

Just realize that there is no right or wrong answer when it comes to choosing a variable, as long as you choose a letter or symbol.

Worksheet 2

Go ahead and write out equations for below 10 statements below:

1. Twelve less than a number is equal to 5.

2. Three times a number, increased by 7, equals 22.

3. The sum of twice a number and 9 is 21.

4. Five more than half of a number is 14.

5. A number divided by 4, then decreased by 3, equals 8.

6. Emma buys three shirts, each costing the same amount. After giving the cashier $50, she gets $5 back. How much does each shirt cost?

7. A recipe calls for twice as much flour as sugar. If you use 4 cups of flour, how much sugar do you need?

8. A car travels at 50 mph. How long does it take to cover 200 miles?

9. Jake saves $20 per week. After some weeks, he has saved $180. How many weeks did he save?

10. A theatre charges $12 per ticket. If a family spends $84, how many tickets did they buy?

Bonus #1 – Scientific Calculator Guide

If you've liked the book so far, you'll love this.

If you haven't used a scientific calculator for a while, you're going to need a bit of a refresher. We're giving you a free guide on what the different symbols on the calculator mean. This guide goes great with the content in this book, and you will definitely find it useful.

Just scan the QR code below to get it.

Solving Basic Algebraic Equations

There are countless ways to solve Algebraic equations, but let's focus on the basics for now.

Before we get to the steps involved, let's look at an important concept called Inverse Operations.

Inverse Operations are basically opposite operations.

For addition, the inverse operation is subtracting the same number. For subtraction, the inverse operation is adding the same number.

For multiplication, the inverse operation is dividing by the same number.

For division, the inverse operation is multiplying by the same number.

So, the inverse operations for:

$+ 5 \rightarrow -5$

$- 4 \rightarrow + 4$

$\times 4 \rightarrow \div 4$

$\div 3 \rightarrow \times 3$

So, when we talk about inverse operations, we are doing the opposite operation on an Algebraic term (constant or variable).

So, now let's have a look at the steps involved using the sample equation below.

x + 7 = 13

Step 1: Move all Variables to One Side Using Inverse Operations

In this case, we have only 1 variable x, which is already on one side, the LHS.

Step 2: Isolate all Constants to the Other Side Using Inverse Operations.

In the next case, move the variable 7 to the other side. To do this, we use the inverse operation of "+7", which is, "-7"

+ 7 → -7

So, we apply -7 to both sides of the equation

x + 7 − 7 = 13 − 7

x + 0 = 6

x = 6

Step 3: Find Unit Value of Variable Using Inverse Operations to Both Sides

We already have the value of one unit of x as 6. So, the answer is:

x = 6

Now, let's look at another example:

$$\frac{4a}{3} = 10$$

'a' is the variable we need to find the value of.

Step 1: Move all Variables to One Side Using Inverse Operations

We only have one instance of the variable 4a/3; and it's already on one side.

Step 2: Isolate all Constants to the Other Side Using Inverse Operations.

We only have one constant 10 and it's already on one side.

Step 3: Find Unit Value of Variable Using Inverse Operations to Both Sides

So, we have: $\frac{4a}{3} = 10$

We need to do two inverse operations to get the unit value of "a".

First, we have the LHS divided by 3. The inverse of dividing by 3 is multiplying by 3.

÷3 → X 3

So, we multiply both sides by 3:

$$\frac{4a}{3} \times 3 = 10 \times 3$$

4a = 30

So, we now have 4 multiplied by a. The inverse of multiplying by 4 is dividing by 4.

× 4 → ÷4

$$\frac{4a}{4} = \frac{30}{4}$$

a = 7.5

So, the value of a is 7.5.

Now, let's look at a third example.

$3x + 7 = 15$

Step 1: Move all Variables to One Side Using Inverse Operations

The variable 3x is already on the left-hand side (LHS), so no changes are needed in this step.

Step 2: Isolate the Constant (7) to the Other Side

The equation is:

$3x + 7 = 15$

Since we want to isolate x, we eliminate the constant 7 using the inverse operation of addition (+7), which is subtraction (-7):

$3x + 7 - 7 = 15 - 7$

$3x + 0 = 8$

$3x = 8$

Step 3: Find the Unit Value of the Variable

The equation now states that 3x equals 8. Since x is multiplied by 3, we perform the inverse operation (division by 3) on both sides:

$$\frac{3x}{3} = \frac{8}{3}$$

x = 2.67

Worksheet 3

Solve the below equations using the strategies in the chapter:

1. x – 12 = 5
2. 3x + 7 = 22

3. $x/2+5=14$ 4. $x/4 - 3 = 8$

5. $3x + 5 = 50$ 6. $2x + 9 = 21$

7. $2s = 4$ 8. $50t = 200$

9. $20w = 180$ 10. $12t = 84$

Verifying Your Answer

To verify a simple algebra answer, follow these steps:

Step 1: Substitute the solution: Replace the variable in the original equation with the value you found.

Step 2: Perform the calculations: Simplify both sides of the equation.

Step 3: Check if both sides are equal: If they are, your solution is correct; if not, recheck your steps.

Now, let's look at a solution for the equation below:

$3x + 7 = 15$

x = 8/3 is the answer as we found out earlier. So, let's go ahead and verify this.

Step 1: Substitute x = 8/3 into 3x + 7 = 15

$3(8/3) + 7 = 15$

Step 2: Perform the calculation on both sides of the equation.

The LHS of Equation = $3(8/3) + 7 = 8 + 7 = 15$

The RHS of Equation = 15

Step 3: Check if both sides are equal

Both sides of the equation are equal to 15.

$LHS = RHS = 15$

So, the answer is correct.

Worksheet 4

Check the proposed answers and verify if they are correct.

1. $x - 8 = 15$

   Proposed Answer: $x = 23$

2. $4x + 5 = 25$

   Proposed Answer: $x = 6$

3. $x/3 - 2 = 7$

   Proposed Answer: $x = 27$

4. $2x + 3 = 19$

   Proposed Answer: $x = 8$

5. $5x - 4 = 21$

   Proposed Answer: $x = 5$

6. $3(x - 2) = 12$

   Proposed Answer: $x = 5$

7. $x/5 + 4 = 10$

    Proposed Answer: x = 30

8. $6x - 2 = 16$

    Proposed Answer: x = 3

9. $10x + 15 = 95$

    Proposed Answer: x = 8

10. $x/2 + 7 = 12$

    Proposed Answer: x = 12

Laws of Algebra

Just as with any other concept, Algebra has a set of laws based on which it operates. These fundamental laws are important to understand as they help build a platform of knowledge we can build on.

1. **Commutative Law**

    If two numbers or variables (or combination of both) are added together then the answer is the same regardless of the order of the numbers or variables. Same is the case with multiplication of the numbers.

    - **For Addition:** $a + b = b + a$

    - **For Multiplication:** $a \times b = b \times a$

Examples:

    - $3 + 5 = 5 + 3 \rightarrow$ both equal 8

    - $4 \times 7 = 7 \times 4 \rightarrow$ both equal 28

Algebra Examples:

    - $4 \times a = a \times 4 \rightarrow$ both equal 4a

2. Associative Law

This law states that the way numbers are grouped does not change the result for addition or multiplication.

- **For Addition:** $(a + b) + c = a + (b + c)$
- **For Multiplication:** $(a \times b) \times c = a \times (b \times c)$

Examples:

- $(2 + 3) + 4 = 2 + (3 + 4) \rightarrow$ both equal 9

- $(5 \times 2) \times 3 = 5 \times (2 \times 3) \rightarrow$ both equal 30

Algebra Examples:

- $(4 \times a) \times 2 = 4 \times (a \times 2) \rightarrow$ both equal 8a

- $(x \times 2) \times a = x \times (2 \times a) \rightarrow$ both equal 2ax

3. Distributive Law

This law states that multiplication distributes over addition or subtraction. In other words, if you multiply the sum of two numbers by a fixed number, it's the same as multiplying each number by the fixed number and adding the two results.

- **Formula:** $a \times (b + c) = (a \times b) + (a \times c)$

Examples:

- $2 \times (3 + 4) = (2 \times 3) + (2 \times 4) \rightarrow$ both equal 14

- $5 \times (6 + 2) = (5 \times 6) + (5 \times 2) \rightarrow$ both equal 40

Algebra Examples:

- $3 \times (a + 5) = (3 \times a) + (3 \times 5) \rightarrow$ both equal $(3a + 15)$

- $5 \times (x + 2) = (5 \times x) + (5 \times 2) \rightarrow$ both equal $(5x + 10)$

4. Identity Law

This law states that adding 0 to a number does not change its value, and multiplying by 1 keeps the number the same.

- **For Addition:** $a + 0 = a$

- **For Multiplication:** $a \times 1 = a$

Examples:

- $7 + 0 = 7$

- $9 \times 1 = 9$

5. Inverse Law

This law states that every number has an opposite (additive inverse) and a reciprocal (multiplicative inverse). This means that any number when added to its additive inverse has a sum of zero; and when multiplied by its reciprocal has a product of 1.

- **For Addition:** $a + (-a) = 0$

- **For Multiplication:** $a \times 1/a = 1$ (for $a \neq 0$)

Examples:

- $8 + (-8) = 0$

- $4 \times \frac{1}{4} = 1$

6. Zero Property of Multiplication

This law states that any number multiplied by 0 is always 0.

- **Formula:** $a \times 0 = 0$

Example:

- $5 \times 0 = 0$

- $0 \times 100 = 0$

Worksheet 5

Fill in the blanks below using the laws of Algebra:

1. $7 + x = x +$ _

2. $5 \times a =$ _ $\times 5$

3. $a + (b + c) = ($_$+$_$) + c$

4. $y \times (a \times 0) =$ _____

5. $y + a + 0 =$ _____

6. $6 \times (x + 5) =$ _____

7. $9 \times 0 =$ _____

8. $Y \times$ _ $= 1$

9. $2 \times (y + 6) =$ ____

10. $3 \times (4 + 5) =$ _____

Index and Power in Algebra

When a number is multiplied by itself a certain number of times, it is raised to that power. For example, if 9 is multiplied by itself 3 times, it is: $9 \times 9 \times 9$

This means that 9 is raised to the power of 3.

$9 \times 9 \times 9 = 9^3$

9 is the **base**; and 3 is the **index or power** of the base.

This works the same way in Algebra. If a variable a is multiplied by itself 5 times, then a is the base and 5 is the exponent or power.

$a \times a \times a \times a \times a = a^5$

Now, just like general algebra, exponents work under certain laws.

Laws of Exponents

1. **Multiplication Rule:** When multiplying numbers with the same base, add the indices.

$a^m \times a^n = a^{m+n}$

Example: $3^2 \times 3^4 = 3^{2+4} = 3^6$

2. **Division Rule:** When dividing numbers with the same base, subtract the indices.

$a^m \div a^n = a^{m-n}$

Example: $5^6 \div 5^2 = 5^{6-2} = 5^4$

3. **Power of a Power Rule:** When raising a power to another power, multiply the indices.

$(a^m)^n = a^{m \times n}$

Example: $(2^3)^2 = 2^{3 \times 2} = 2^6$

4. **Zero Index Rule:** Any number raised to the power of zero is always 1.

$a^0 = 1$

Example: $7^0 = 1$

5. **Negative Index Rule:** A negative exponent means you take the reciprocal (flip the fraction).

$a^{-n} = 1 / a^n$

Example: $3^{-2} = 1/3^2 = 1/9$

Worksheet 6

Simplify below exponents in space below

1. 2^3

2. 5^2

3. 10^4

4. Write 1000 as a power of 10.

5. What is the value of 7^0?

6. $3^2 \times 3^4$

7. $4^5 \times 4^3$

8. $6^7 \div 6^3$

9. $9^5 \div 9^2$

10. $2^4 \times 2^2$

11. $(2^3)^2$

12. $(5^4)^2$

13. $(3^2)^3$

14. $(7^3)^2$

15. 8^{-2}

16. 4^{-3}

17. Express as a fraction: 10^{-1}

18. Express as a fraction: 3^{-2}

19. $(2^3 \times 2^{-1}) \div 2^2$

20. $(5^4 \times 5^{-2}) \div 5^1$

Different Types of Algebraic Equations

Now that we know the basics of Algebra and Algebraic Equations and know how to solve simple equations, let's look at the different types of Algebraic Equations and also look at what we'll cover in this book.

Equations can be divided into different types based on either

A. Number of variables in the Equations
B. Highest Power Variable of the Equation

Group A:

Equations in this group are either single or multi-variable equations.

Single Variable Equations

These equations have only one variable.

Example: $3x + 5 = 11$

x is a single variable in the equation.

Multi-Variable Equations

These equations contain two or more variables.

Example:

$x + 2y = 7$

x, y are two variables in this multi-variable equation.

$x + y + z = 9$

x, y, z are three variables in this multi-variable equation.

Group B:

In this group, equations are grouped by the highest power variable of the equation.

The highest power variable is also called the order of the equation.

For example, in Equation below:

$x^4 + x^3 + x + 1 = 5$

x^4 is the variable with the highest power, and the highest power or order is 4.

Linear Equations

These are first-order equations where the highest power of the variable(s) is 1.

Single Variable: $2x - 5 = 7$

Multi-Variable: $3x + 4y = 12$

Quadratic Equations

These are second-order equations where the highest power of the variable is 2.

Example: $x^2 - 4x + 4 = 0$

Higher-Order Equations

These involve variables with powers greater than 2.

Example: $x^3 + 2x^2 - 5x + 7 = 0$ (Cubic Equation)

Example: $x^4 + 2x^2 - 5x + 7 = 0$ (4th order Equation)

Other Types of Equations

Besides the above equations, there are a few other types below:

Exponential Equations

These have variables in the exponent.

Example: $2^x = 16$

Logarithmic Equations

These involve logarithms of variables.

Example: $3\log_2(x)=3$

Rational Equations

These contain variables in fractions. These are also called Fractional Equations.

Example: $\dfrac{x+2}{x+3} = 2$

In this book we'll focus on single and double variable equations of the first and second order. We'll also have examples of basic rational, logarithmic and exponential equations.

49

Worksheet 7

What type of equations are each of the below equations?

1. $3x + 7 = 10$

2. $x^2 - 5x + 6 = 0$

3. $2x + 3y = 12$

4. $\dfrac{x+2}{x-3} = 4$

5. $x^3 - x^2 + 5x + 2 = 0$

6. $5^x = 25$

7. $\log_3 (x) = 125$

8. $7x - 3 = 4x + 2$

9. $x^4 - 2x^2 + x - 1 = 0$

10. $2x - y = 5$

11. $6x^2 - 11x + 2 = 10$

12. $4^x = 64$

13. $\log (x + 3) = 2$

14. $\dfrac{3x}{x+5} = 7$

15. $x^5 + 3x^3 + 2x^2 + 5 = 0$

Graphing Linear Equations

Graphing Points on an Axes

On a graph, the y-axis is the vertical line that goes from top to bottom. The x-axis is the horizontal line that goes from left to right. The x-axis and y-axis intersect at a point called the origin or (0,0).

Any point above the origin is positive on the Y-axis. It is represented by +Y on the graph. Any point below the origin is negative on the Y-axis. It is represented by -Y on the graph.

Any point to the right of the origin is positive on the X-axis. It is represented by +X on the graph. Any point to the left of the origin is negative on the X-axis. It is represented by -X on the graph.

Have a look at the graph below to see what it looks like.

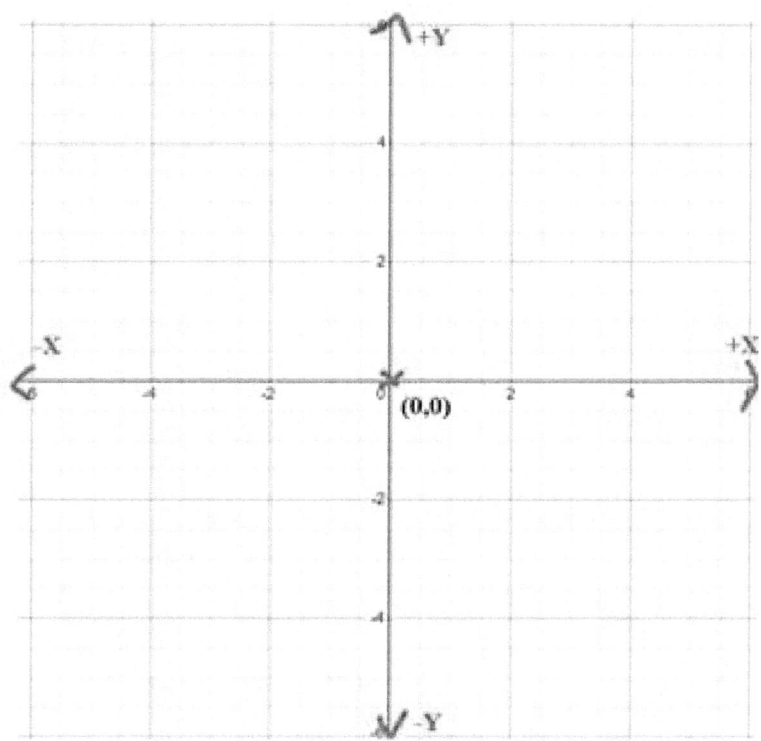

Now, let's plot some points on the graph. The first point is (2,3). The first co-ordinate is the x-value. Since it is positive it is on the right-hand side of the origin. +2 means it is 2 units to the right of the origin.

The second co-ordinate is the y-value. Since it is positive it is above the origin. +3 means it is 3 units to the top of the origin.

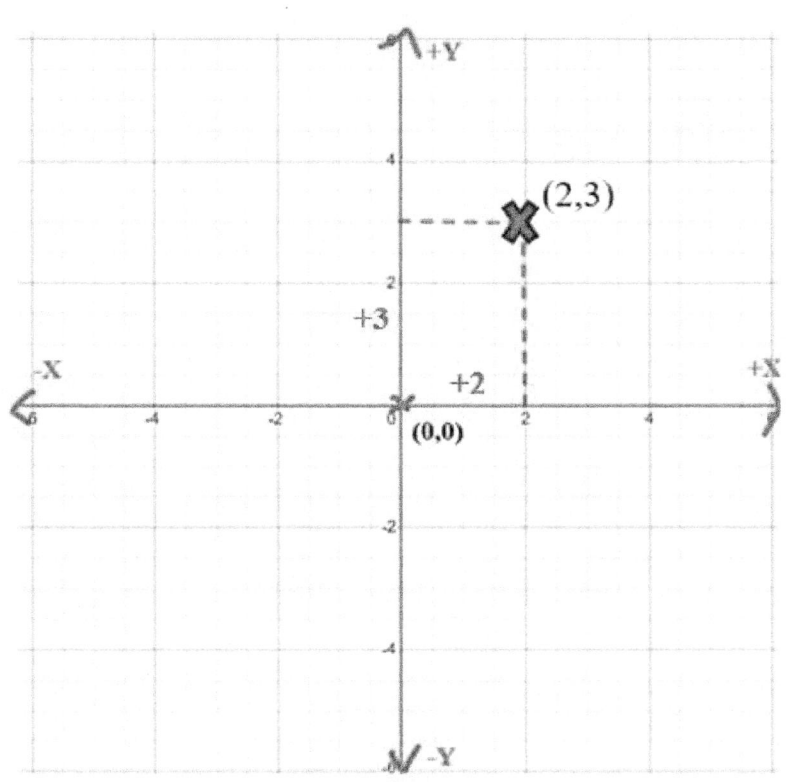

Now, let's look at another point. Let's look at (-2,4). The first co-ordinate is the x-value. Since it is negative it is on the left-hand side of the origin. -2 means it is 2 units to the left of the origin.

The second co-ordinate is the y-value. Since it is positive it is above the origin. +4 means it is 4 units to the top of the origin.

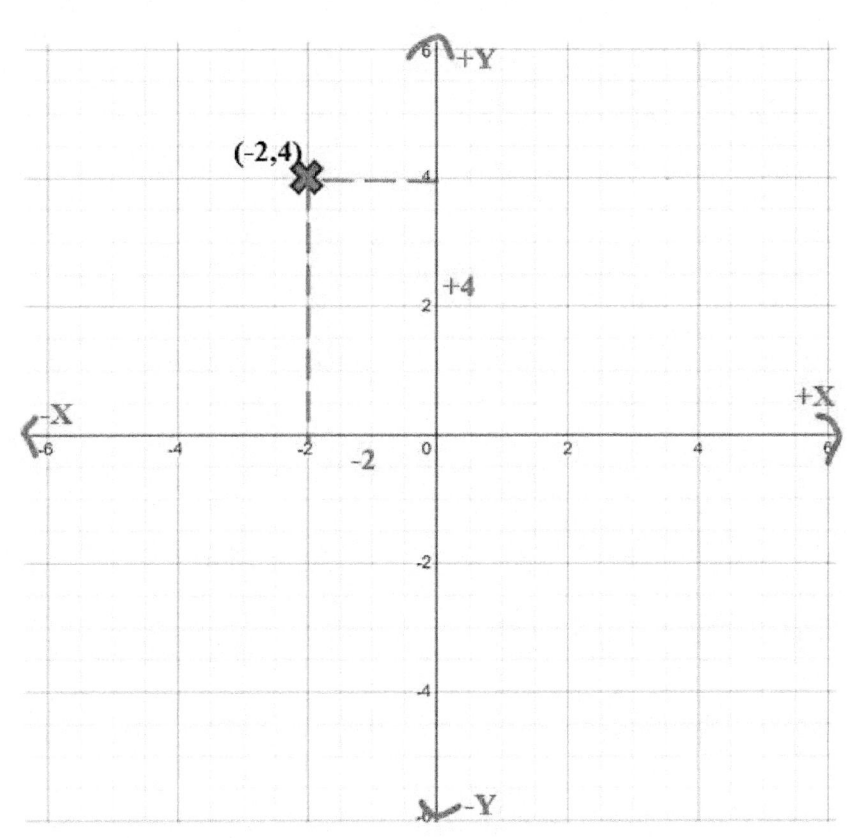

Worksheet 8

Plot the following 8 points on the graph below.

  a. (2,2)   b. (-7,3)   c. (3, -2)   d. (-2, -2)

  e. (5,5)   f. (4, -3)   g. (0,4)   h. (0,0)

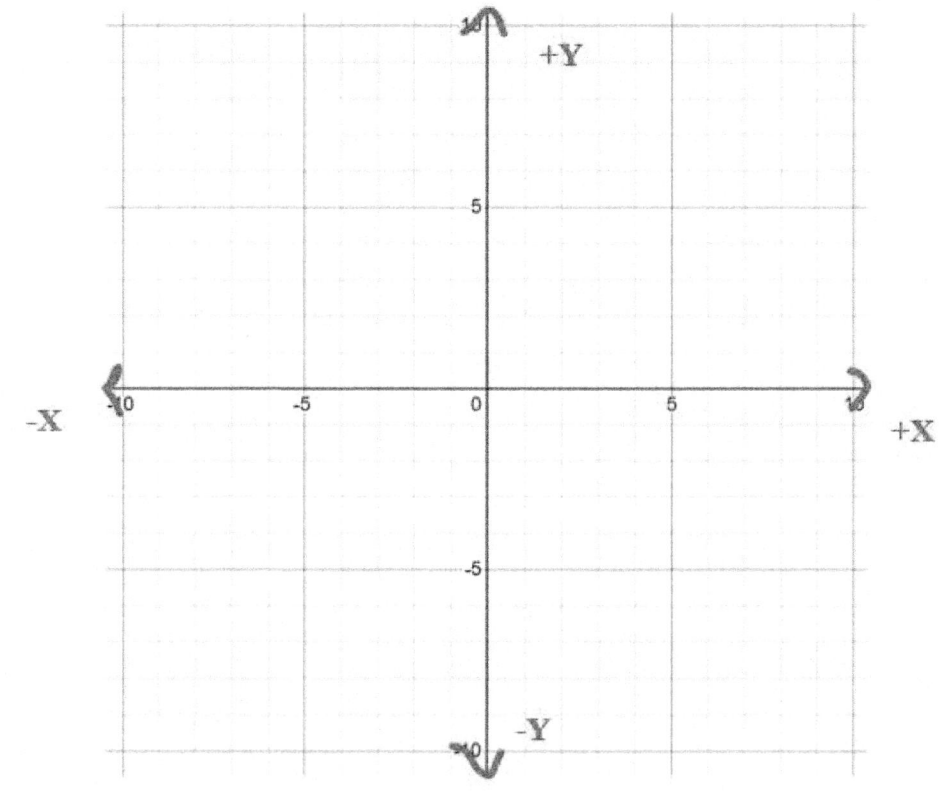

56

Lines On a Graph

A line is just the shortest distance between two points on a graph.

The line represents the two points, and also the possible points between the two points. Lines are represented by linear equation. All points on a line are represented by this equation.

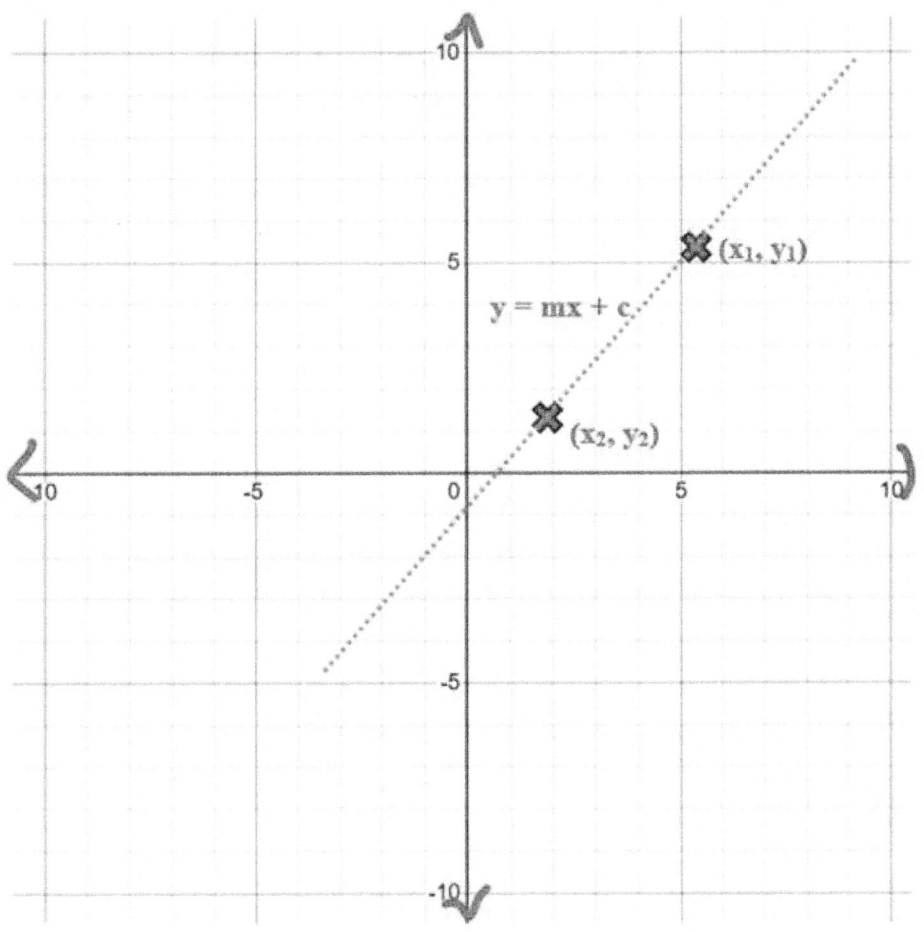

Why Represent Linear Equations on a Graph

A linear equation (e.g., $y = 2x + 3$) shows a relationship between two variables. This relationship that forms a straight line when plotted on a coordinate plane (x-axis horizontal, y-axis vertical) as shown below.

Graphing turns an equation into a picture, making it easier to see patterns—like how y changes with x—and solve real-world problems, like predicting costs or distances over time. For example, $y = 2x + 3$ could represent earning $2 per hour plus a $3 bonus. Graphing shows your total pay (y) for hours worked (x)—a straight line rising steadily.

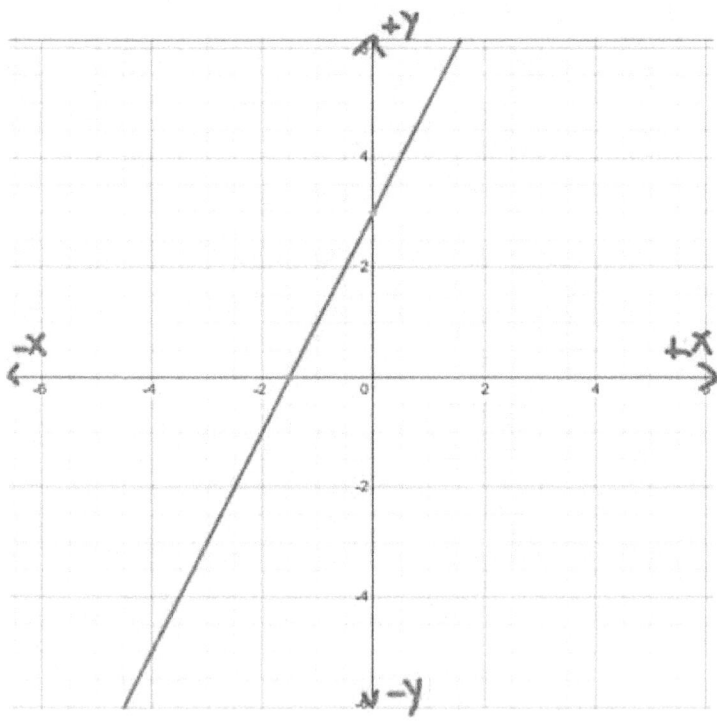

Equation of a Line

The equation of a line is given by the formula:

$$y = mx + c$$

x, y are the unknown variables.

m is the change in the value of y per unit change in x. m is also called the gradient or slope of the line.

c is the point at which the line intercepts the y- axis.

All points on the line are the variables (x,y).

Finding Points on a Line

As we mentioned earlier, a linear equation represents all points on a line. So, if we have the equation $y = 2x + 3$, does the point (2,3) or (1,5) lie on the line?

To figure this out, we just plug the points in the equation:

The first point is (2,3).

Plug in $x = 2$, $y = 3$

So, $2x + 3 = 2(2) + 3 = 4 + 3 = 7$

But, $y \neq 7$, $y = 3$.

So, the point does not line on the line as it does not fit the equation.

Now, let's look at a second point (1,5).

Plug in x = 1, y = 5

So, 2x + 3 = 2(1) + 3 = 2 + 3 = 5

We also know that, y = 5.

So, the point does lie on the equation.

All right. Now what if we're given one of the co-ordinates and we need to find the other? We just plug it in the equation and find the other co-ordinate.

So, for the same equation y = 2x + 1, what if we're given x = 3? What is the value of y given x = 3?

y = 2(3) + 1 = 7.

So, when x = 1, y = 7 and the point (1,7) lies on the line.

Let's look at another example. For the same equation y = 2x + 1, what is the value of x when y = 3.

We plus y = 3 into the equation.

3 = 2x + 1

Subtract 1 from both sides..

$3 - 1 = 2x + 1 - 1$

$2 = 2x$

Divide both sides by 2

$1 = x$ OR

$x = 1$

So, when $y = 3$, $x = 1$ on the point (1,3).

Worksheet 9

1. Which of the following points lie on the line $y = 5x + 1$?

   a. (2,3)   b. (1,6)   c. (2,11)   d. (1,1)   e. (-1, -4)

2. Find the values of the coordinates in the equations below

a. Find y value for $x = 1$ in $y = 2x - 1$

b. Find y value for $x = 2$ in $y = 3x + 1$

c. Find y value for $x = 0$ in $y = 2x$

d. Find x value for $y = 2$ in $y = 9x - 1$

e. Find x value for $y = 1$ in $y = 3x + 2$

Find the Slope

Now that we know how a linear equation works for points, let's look at how it defines the actual line on a graph.

As we mentioned earlier, the equation of a line is:

y = mx + c

'm' is the slope of the gradient of the line. It is the rate at which y changes given a change in value of x.

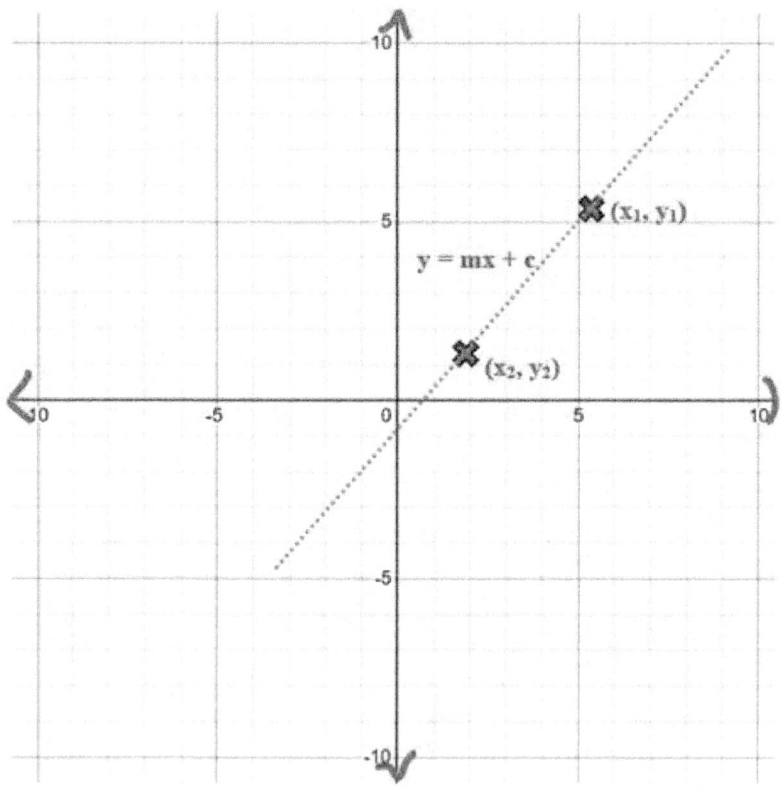

For example, in the equation $y = 2x + 1$.

2 is the coefficient of x, so $m = 2$.

The line $y = 2x + 1$ has a slope or gradient of 2.

So, for every time the value of x increases by 1, y increases by 2 and so on.

Now, let's look at a line $y = 7x + 2$

The slope is 7 since 7 is the coefficient of x.

Now, what about this equation below?

$2y = 3x + 2$

Is the slope 3? 3 is the coefficient of x.

No.

We need to get in the correct form $y = mx + c$

So, what we need to do is divide both sides of 2 so we get the equation in the correct form.

$2y = 3x + 2$

Dividing both sides by 2, we get:

$y = 1.5x + 2$

So, m = 1.5 for the line $y = 1.5x + 2$

Now, what if we don't know the equation of the line?

We're only given two points (x_1, y_1) and (x_2, y_2). These points lie on the line.

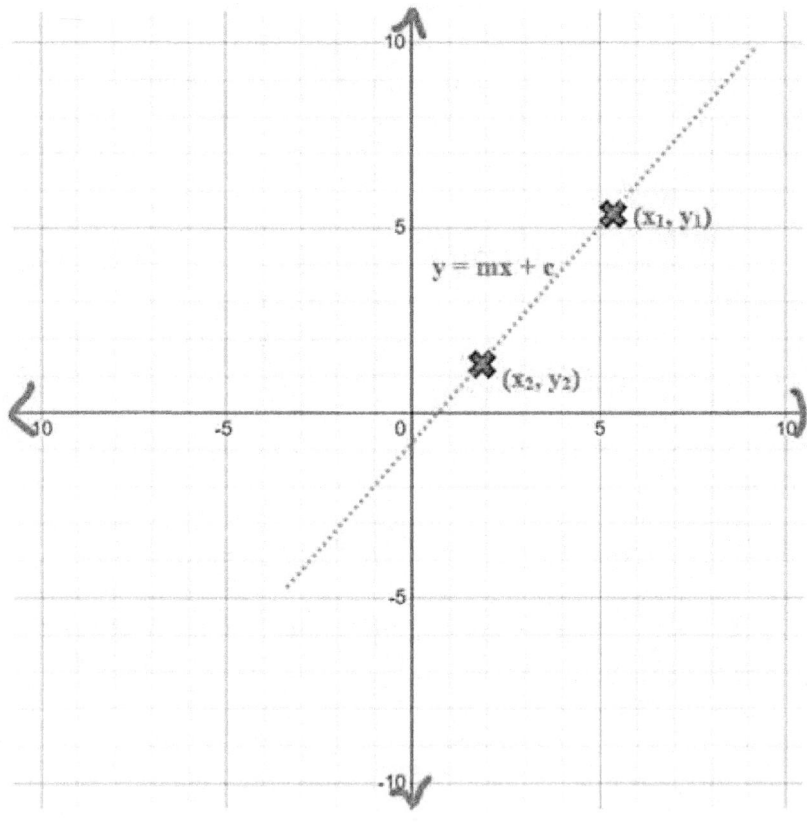

So, the slope m is given by the change in y-coordinates divided by the change in x-coordinates.

$$m = \frac{y_2 - y_1}{x_2 - x_1}$$

Now, let's look at a simple example. What is the slope of the line that passes through the points (1,2) and (4,5)?

We just simply use the equation:

$$m = \frac{y_2 - y_1}{x_2 - x_1}$$

$$m = \frac{5 - 2}{4 - 1} = \frac{3}{3} = 1$$

So, the slope is 1.

Worksheet 10

1. What is the slope of the below equations?

    a. $y = 3x + 1$

    b. $3y = 6x + 4$

    c. $2x + 3y = 1$

    d. $y = 7x + 4$

    e. $4x + 2y = 9$

2. What is the slope of the line that passes through two points below?

   a. (1,2) and (4,5)

   b. (7,8) and (9,11)

   c. (3,4) and (7,8)

   d. (3,5) and (8,9)

   e. (1,1) and (2,5)

Find the y-Intercept of a Line

Now that we understand how to find the slope of a line, let's look at the second element of below equation of a line.

$y = mx + c$

c is the y-intercept or the point at which the line intercepts the y-axis.

In the diagram below, the blue point is the y-intercept or value of c.

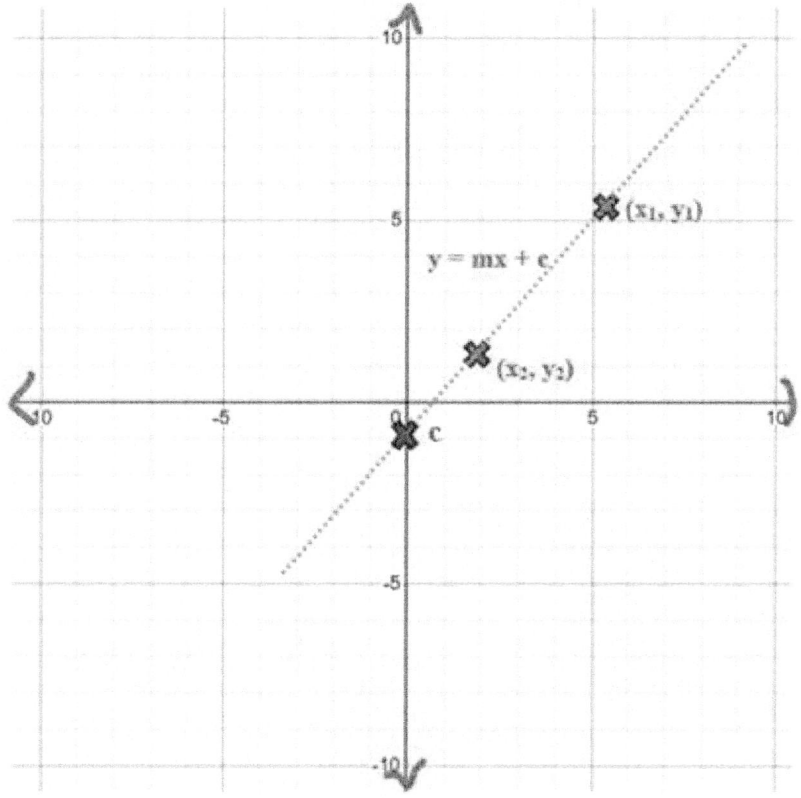

c can be written as (0, c) where the x-ordinate is 0.

So, we can easily find the y-intercept by plugging in x =0. For example, in the equation y = 2x + 1, we can find the y-intercept by plugging in x = 0.

y = 2(0) + 1

y = 1.

So, the y-intercept is 1 or c = 1.

Worksheet 11

Find the y-intercept of below equations:

   a. y = 3x + 4

   b. y = −2x + 7

   c. y = 12x − 5

   d. y = −5xy

e. $y = 7x + 0$

f. $y = -34x + 2$

g. $y = 10x - 3$

h. $y = 6 - x$

i. $y = 23x + 9$

j. $y = -8x - 1$

Plotting a Line on a Graph

Now, what if we're given an equation.

$y = 3x + 1$

How do we plot it on a graph?

The best way is to mark two points on the graph and then draw a line through the two points.

So, for above equation let's pick the first point. I always prefer the first point to be the y-intercept.

So, let's set $x = 0$ for the first point. We get:

$y = 3(0) + 1 = 1$

So, the first point is (0,1). Let's mark that on the line.

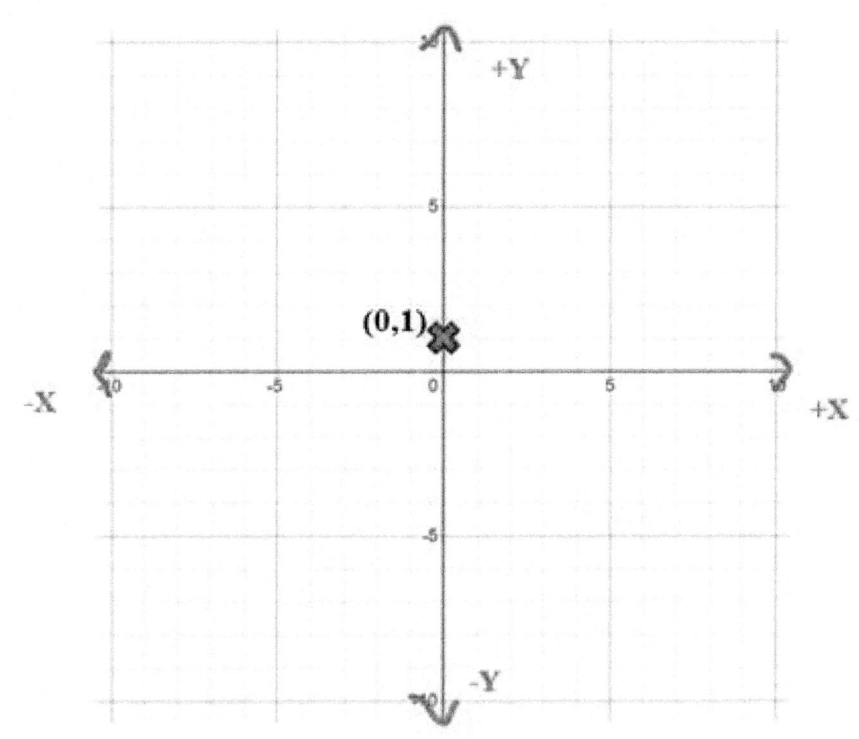

Now, let's pick the second point. Let's set x = 1. Then we get:

y = 3(1) + 1 = 4.

So, the second point is (1,4)

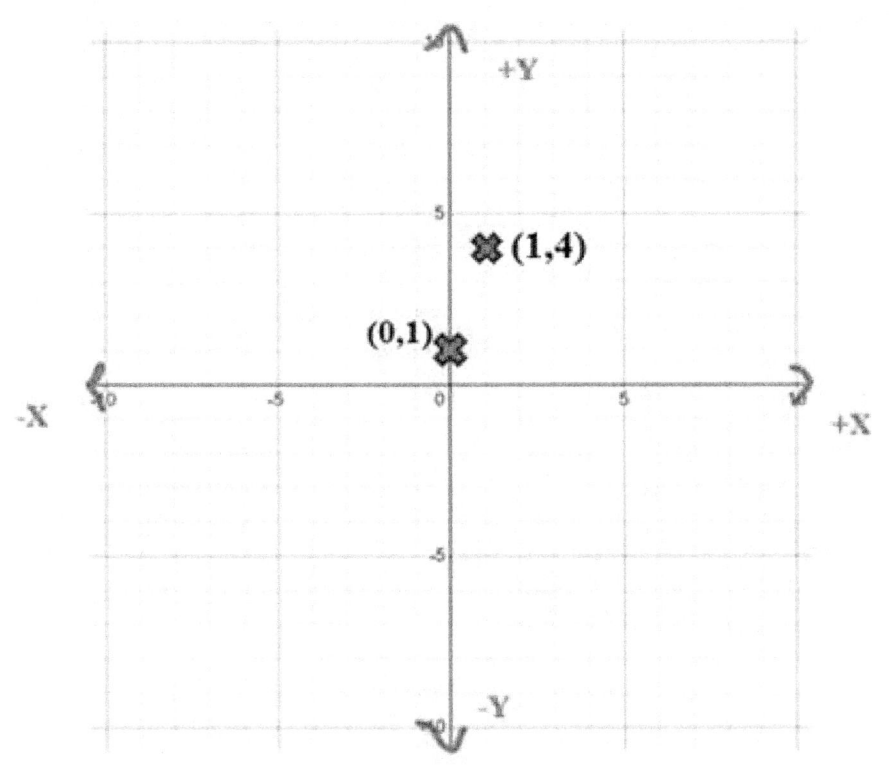

Now, we can just draw a line through the two points, and we get the line we are looking for.

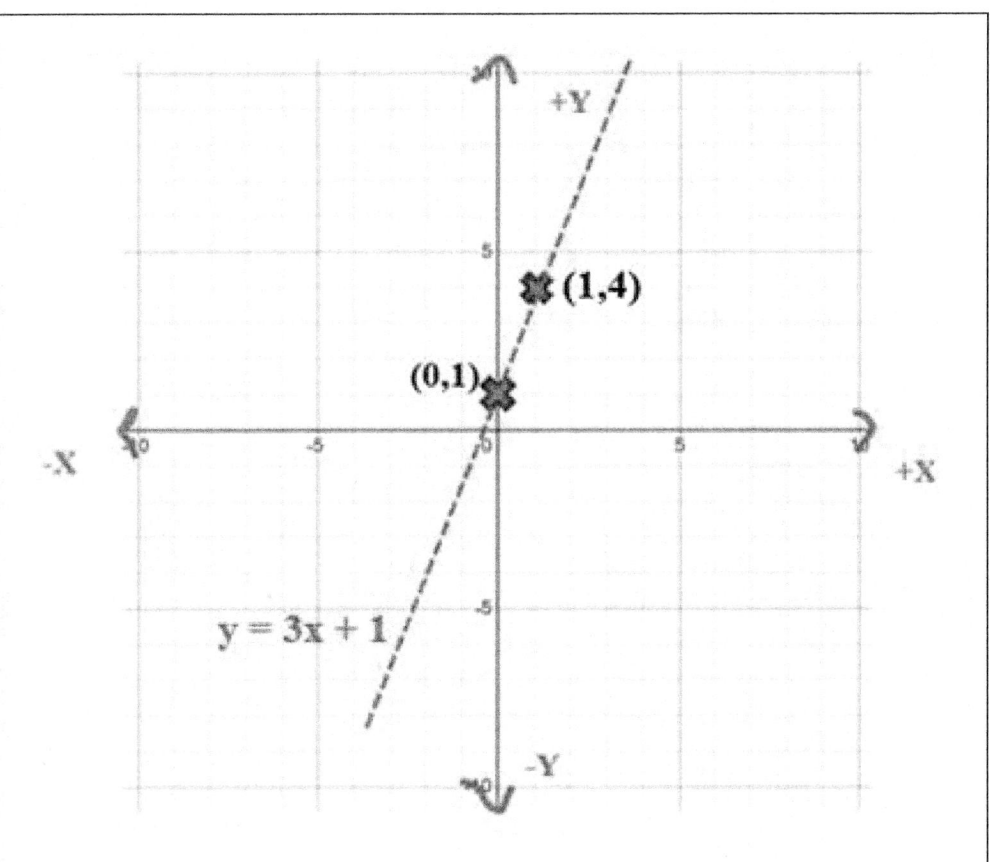

Worksheet 12

Draw the lines for below equations on the graph below.

a. $y = 2x + 1$

b. $y = 5x + 2$

c. $2y = 4x + 4$

d. $x + y = 2$

e. $2x + 5y = 10$

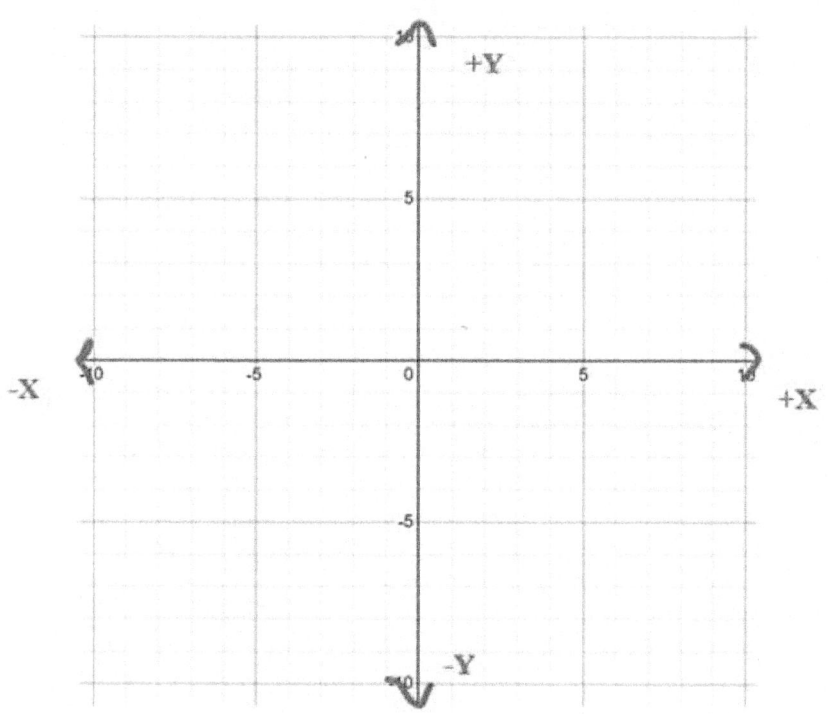

74

Finding Equation of a Line

Now, what if we have a line and we don't know the equation of a line? There are three different ways to find the equation depending on what information we have available.

a. Given Two Points On The Line

If we have the line on the graph, we can spot two points on the graph. And we can find the equation of the line using the slope. The two points on the graph below are (x_1, y_1) and (x_2, y_2)

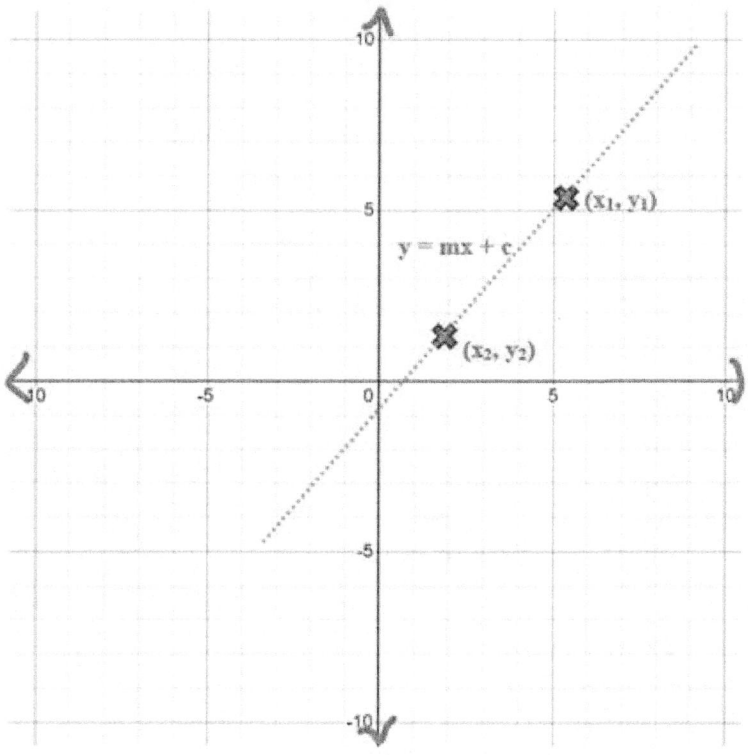

The first step is to find the slope using the two points below.

$$m = \frac{y_2 - y_1}{x_2 - x_1}$$

The second step is to find the y-intercept by setting x = 0

y = m(0) + c = c

So, the y-intercept is c.

We can plug in the values of y-intercept and slope to get the equation.

For example, let's look at below graph. We identify two points (2,7) and (4,11).

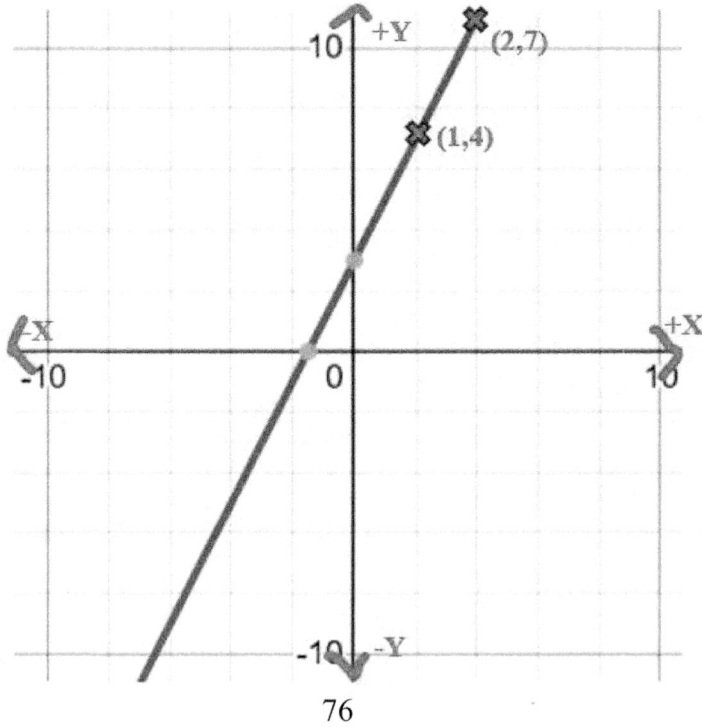

So, we complete the first step by finding the slope

$$m = \frac{y_2 - y_1}{x_2 - x_1} = \frac{11 - 7}{4 - 2} = \frac{4}{2} = 2$$

So, the slope is 2.

Now, we can find the y-intercept from the graph.

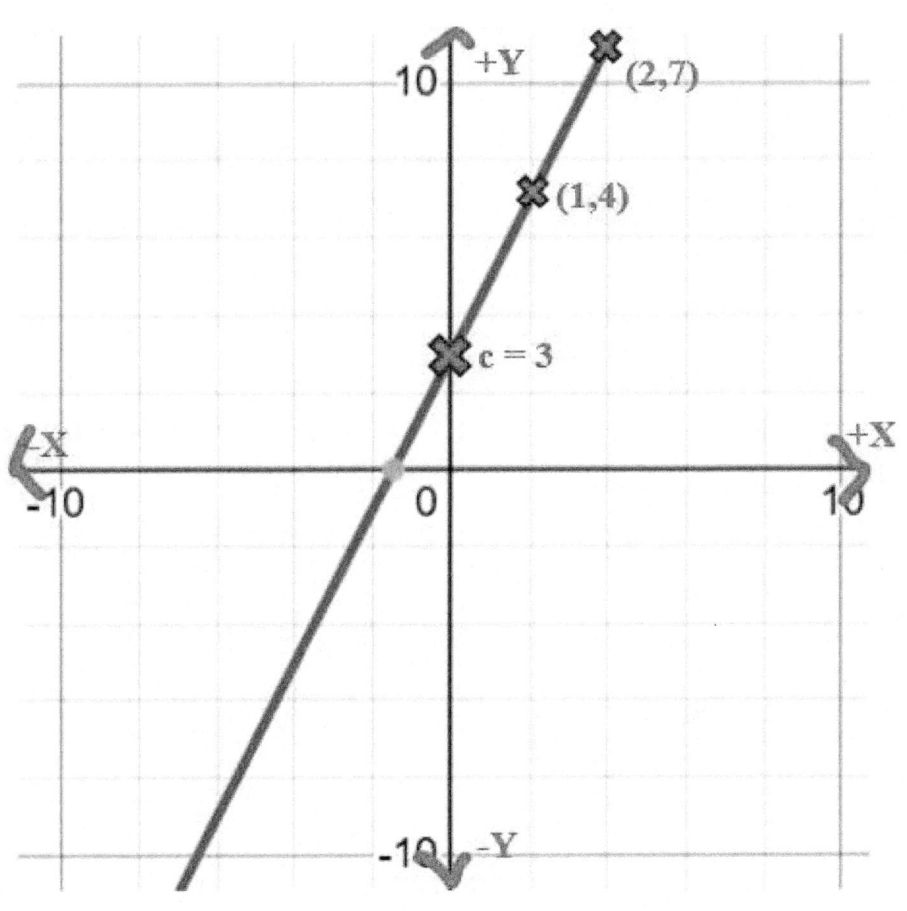

The y-intercept from the graph is 3 from the graph. So, the value of c is 3.

So, when we plug the values of m and c into the equation $y = mx + c$, we get:

$y = 2x + 3$.

b. Given Slope and a Point

Now, what if we're only given one point? We need an additional piece of information such as the slope to solve the equation.

$y = mx + c$

m is the slope and (x_1, y_1) is the given point.

We just plug in the value of (x_1, y_1) and m into the equation and find the value of the y-intercept c.

For example, we're given a point (-1, 4) as the point and a slope of 2.

So, we plug these into the equation $y = mx + c$

$y = 4$, $x = -1$ and $m = 2$

$4 = -1(2) + c$

$4 = -2 + c$

We add + 2 to both sides of the equation

$4 + 2 = -2 + 2 + c$

$c = 6$

When we plug back in m = 2 and c = 6 back into the equation, we get:

$y = 2x + 6$

c. Given One Point and y-Intercept

Now, what if we're only given one point and y-intercept? We just plug these back into the equation.

$y = mx + c$

m is the slope and (x_1, y_1) is the given point.

We just plug in the value of (x_1, y_1) and m into the equation and find the value of the slope m.

For example, we're given a point (2, -6) as the point and a y-intercept of 5.

So, we plug these into the equation $y = mx + c$

$y = -6$, $x = 2$ and $c = 5$

$-6 = m(2) + 5$

$-6 = 2m + 5$

We subtract 5 from both sides of the equation

-6 -5= 2m + 5 - 5

-11 = 2m

Divide both sides of the equation by 2, we get:

$$m = -\frac{11}{2}$$

When we plug back in m = -11/2 and c = 5 back into the equation, we get:

$$y = (-\frac{11}{2})x + 5$$

OR

2y = -11x + 5 (Multiplying both sides by 2 to get rid of the fraction)

Worksheet 13

Find the slope using the formula: $m = (y_2 - y_1) / (x_2 - x_1)$. Then, use one point and the slope to find the y-intercept and write the full equation $y = mx + c$.

1. Find the slope between the two points:

   a. $(1, 3)$ and $(3, 7)$

   b. $(2, 4)$ and $(6, 12)$

2. Using your answer from 1, find the y-intercept.

3.   Given: Slope = 3, Point = $(2, 8)$.

      Plug into the equation and find the y-intercept (c).

4. Write the full equation of the line above.

5. Given: Slope = -2, Point = (-1, 1).
   Find the y-intercept (c).

6. Write the full equation of the line using your answer to #6.

7. Point = (3, -1), y-intercept = 5.
   Find the slope (m).

8. Write the full equation of the line.

9. Point = (-4, 2), y-intercept = -6.

   Write the full equation of the line.

10. Point = (3, 9), y-intercept = 2.

   Write the full equation of the line.

Bonus #2 – Formula Worksheet

If you've enjoyed this book so far, you'll love this.

It's a list of formulas used in general mathematics. It's not only a refresher, it's also a cheat sheet for when you need to use math concepts and you don't want to search for the right page in the right book.

Just scan the QR code below to get it.

Solving Fractional Equations

What is a Fractional Algebra Equation?

A **fractional algebra equation** (also known as a *rational equation*) is an equation that contains one or more algebraic fractions—fractions that include variables in the numerator, denominator, or both.

Example:

$$\frac{x+2}{3} = \frac{2x-1}{5}$$

These equations often involve cross-multiplying or finding a common denominator. The goal is to get it into the form of a linear Algebraic Equation. Then we can solve it just like we've done in the rest of the book.

Steps to Solving a Simple Fractional Equation

1. **Cross Multiply:**

   a. Multiply both sides by denominator on LHS. This eliminates the LHS denominator.

   b. Multiply both sides by denominator on RHS. This eliminates the RHS denominator.

2. **Simplify the Resulting Equation:**

   Distribute and combine like terms as needed.

3. **Solve the Equation:**

   Use standard algebraic techniques (isolate the variable, etc.).

Example:

Solve:

$$\frac{x + 2}{3} = \frac{2x - 1}{5}$$

Solution:

Step 1:

  a. LHS denominator is 3. Multiply both sides by 3.

$$\left(\frac{x + 2}{3}\right)3 = \left(\frac{2x - 1}{5}\right)3$$

$$\left(\frac{x + 2}{\cancel{3}}\right)\cancel{3} = \left(\frac{3(2x - 1)}{5}\right)$$

$$x + 2 = \left(\frac{3(2x - 1)}{5}\right)$$

  b. RHS denominator is 5. Multiply both sides by 5.

$$5(x + 2) = \left(\frac{3(2x-1)}{5}\right)5$$

$$5(x + 2) = \left(\frac{3(2x-1)}{\cancel{5}}\right)\cancel{5}$$

$$5(x + 2) = 3(2x - 1)$$

Step 2:

5x + 10 = 6x - 3

Add 3 to both sides

5x + 10 + 3 = 6x – 3 + 3

5x + 13 = 6x

Subtract 5x from both sides

5x + 13 – 5x = 6x – 5x

13 = x

x = 13 is the answer.

Worksheet 14

Solve the following:

1. $\dfrac{x+3}{4} = \dfrac{3x-2}{7}$

2. $\dfrac{2x-1}{5} = \dfrac{x+4}{2}$

3. $\dfrac{x-5}{3} = \dfrac{4x+1}{6}$

4. $\dfrac{3x+2}{8} = \dfrac{x-3}{4}$

5. $\dfrac{x+1}{2} = \dfrac{5x-3}{9}$

Equation with Multiple Fractions

Equations with **more than two** fractional terms follow the same steps as above, with one extra step in between. The two fractions need to be combined into one single fraction on both LHS and RHS.

Example:

$$\frac{x}{2} + \frac{1}{3} = 2x + \frac{x}{3}$$

In this case we have two fractions on the LHS and the RHS. We need to get both the LHS and RHS under a common denominator. This common denominator is the LCM or Lowest Common Multiple.

89

So, how do we do this? We take the two denominators and find their multiples.

$$\frac{x}{2} + \frac{1}{3}$$

The two denominators are 2 and 3. So, let's go ahead and find their multiples.

2 X 1 = 2	3 X 1 = 3
2 X 2 = 4	**3 X 2 = 6**
2 X 3 = 6	3 X 3 = 9
2 X 4 = 8	3 X 4 = 12

So, what we need to figure out is the lowest multiple that is common to both.

In this case, **6** is the lowest multiple.

So, we need to get both denominators under **6**.

To achieve this, we need to multiply numerator and denominator of first fraction by 3. This ensures that the denominator is 6.

And for the second fraction, we multiply numerator and denominator by 2.

$$\frac{x}{2} + \frac{1}{3} = \frac{x \times 3}{2 \times 3} + \frac{1 \times 2}{3 \times 2}$$

90

And then we can combine two equations into one since they have the same denominator.

$$\frac{x \times 3}{2 \times 3} + \frac{1 \times 2}{3 \times 2} = \frac{3x}{6} + \frac{2}{6} = \frac{3x + 2}{6}$$

Now, let's do the same with RHS fractions. 2x can be written as $\frac{2x}{1}$

$$2x + \frac{x}{3} = \frac{2x}{1} + \frac{x}{3}$$

The LCM of 1 and 3 is 3. So common denominator is 3.

1 X 1 = 1	**3 X 1 = 3**
1 X 2 = 2	3 X 2 = 6
1 X 3 = 3	3 X 3 = 9
1 X 4 = 4	3 X 4 = 12

$$\frac{2x}{1} + \frac{x}{3} = \frac{2x \times 3}{1 \times 3} + \frac{x \times 2}{3 \times 1} = \frac{6x}{3} + \frac{2x}{3} = \frac{6x+2x}{3} = \frac{8x}{3}$$

So, our equation now becomes:

$$\frac{3x + 2}{6} = \frac{8x}{3}$$

Now, we solve this as we did for simple fractional equations earlier:

We multiply both sides of the equation by LHS denominator 6.

$$\left(\frac{3x + 2}{6}\right)6 = \left(\frac{8x}{3}\right)6$$

$$\left(\frac{3x + 2}{\cancel{6}}\right)\cancel{6} = \left(\frac{8x}{\cancel{3}}\right)\cancel{6}2$$

$$3x + 2 = 16x$$

Subtract 3x from both sides

$$3x + 2 - 3x = 16x - 3x$$

$$2 = 13x$$

$$x = \frac{2}{13}$$

The process seems a little long, but once we get used to solving steps like the LCM and cross-multiplication, it gets a lot easier.

Let's look at another example.

$$2x + \frac{1}{5} = 4x$$

Since we don't have multiple fractions on RHS, we only need to find LCM for LHS denominators.

The LHS fraction is:

$$2x + \frac{1}{5} = \frac{2x}{1} + \frac{1}{5}$$

The LCM of 1 and 5 is **5**. Now, we work to get a common denominator of 5.

$$\frac{2x}{1} + \frac{1}{5} = \frac{2x \times 5}{1 \times 5} + \frac{1 \times 1}{5 \times 1} = \frac{10x}{5} + \frac{1}{5} = \frac{10x + 1}{5}$$

So, the equation now becomes:

$$\frac{10x + 1}{5} = 4x$$

So, we solve it as before again:

Multiply both sides by 5.

$$\left(\frac{10x + 1}{5}\right)5 = 4x \times 5$$

$$10x + 1 = 20x$$

Subtract 10x from both sides:

$$10x + 1 - 10x = 20x - 10x$$

$$1 = 10x$$

$$x = \frac{1}{10}$$

Equation with Variable in Denominator

Special care is needed when **the variable is in the denominator,** because not all values of the variable are allowed.

Example:

$$\frac{5}{x} + 3 = 2$$

In this case, we cannot have a fraction where the denominator is zero, so we cannot have x = 0. So, $x \neq 0$.

Once we know that it cannot be x = 0, we go ahead and solve the problem as before taking the LCM of the two fractions on the LHS.

$$\frac{5}{x} + 3 = \frac{5}{x} + \frac{3}{1}$$

The LCM of x and 1 is x.

$$\frac{5}{x} + \frac{3}{1} = \frac{5 \times 1}{x \times 1} + \frac{3 \times x}{1 \times x} = \frac{5 + 3x}{x}$$

So, the equation now becomes:

$$\frac{5 + 3x}{x} = 2$$

Multiply both sides by x,

$$\left(\frac{5 + 3x}{x}\right) x = 2x$$

$$5 + 3x = 2x$$

Subtract 2x from both sides:

Subtract 5 from both sides:

$$5 + 3x - 5 = 2x - 5$$

$$3x = 2x - 5$$

Subtract 2x from both sides

$$3x - 2x = 2x - 2x - 5$$

$$x = -5$$

So, the answer is x = -5

Now, let's look at another example:

$$\frac{3}{2x + 3} = \frac{1}{3}$$

In this case, we don't want LHS denominator to be zero. So:

2x + 3 ≠ 0

Solving for x, we get x ≠ -1.5. So, we cannot have x = 1.5

Now, let's go ahead and solve this. In this case, we don't need to take an LCM since we have only 1 fraction on each side of the equation.

So, we multiply both sides by (2x+3).

$$(\frac{3}{2x + 3})(2x + 3) = (\frac{1}{3})(2x + 3)$$

$$(\frac{3}{\cancel{2x + 3}})(\cancel{2x + 3}) = (\frac{2x + 3}{3})$$

$$3 = (\frac{1}{3})(2x + 3)$$

Now, multiply both sides by 3.

$$3 \, X \, 3 = (\frac{1}{3})3(2x + 3)$$

$$9 = 2x + 3$$

Subtract 3 from both sides

$$9 - 3 = 2x + 3 - 3$$

$$2x = 6$$

Divide both sides by 2

We get x = 3 as the answer.

Worksheet 15

1. Find the LCM of below numbers:

    a. $2,3 =$ ____   **b.** $4,5=$ ____   **c.** $1,2=$ ____

    d. $7,3 =$ ____   **e.** $2,4=$ ____   **f.** $2,5=$ ____

2. Combine the two fractions under a single fraction:

    a. $\dfrac{1}{3} + \dfrac{1}{2}$ b. $\dfrac{1}{5} + \dfrac{x}{2}$

    c. $\dfrac{x}{2} + \dfrac{5}{3}$ d. $\dfrac{x}{2} + \dfrac{x}{7}$

    e. $\dfrac{2x}{5} + \dfrac{3x}{4}$ f. $\dfrac{2}{x} + \dfrac{3}{2}$

3. Solve the following equations below:

    a. $7 + \dfrac{3}{x} = 2$

    b. $\dfrac{5}{x} + 3 = \dfrac{3}{2x}$

    c. $7 + \dfrac{x}{3} = 2$

d. $\dfrac{x}{5} + \dfrac{1}{4} = x + \dfrac{2x}{5}$

e. $\dfrac{x}{4} + \dfrac{5}{2} = 3$

f. $\dfrac{x+2}{3} = \dfrac{2x-1}{4}$

g. $3 + \dfrac{1}{x} = \dfrac{4}{5}$

h. $\dfrac{2x}{3} + \dfrac{x}{2} = x + \dfrac{1}{6}$

i. $\dfrac{1}{x} + \dfrac{1}{2} = \dfrac{5}{6}$

j. $\dfrac{4}{x} + \dfrac{2}{3} = \dfrac{5}{6}$

Solving Simultaneous Equations

What are Simultaneous Equations?

Simultaneous equations are a set of two or more equations that share the same set of variables. The goal is to find values for all the variables that make all the equations true. Typically, for two variables (e.g., x and y), two equations are needed to find a unique solution, assuming the equations are independent and consistent.

For example, look at the two equations below:

2x + y = 8

x − y = 1

The above two equations are a set of simultaneous equations with two variables x and y.

The solution is a pair of values for x and y that satisfies both equations. For the above set of equations **x = 3** and **y = 2** makes both equations true. This is written as a solution of **(3,2)** for the set of equations.

Simultaneous equations can be linear (where variables are raised to the power of 1) or non-linear (involving higher powers or other functions). In this workbook, we focus on linear simultaneous equations, which can be solved algebraically or graphically.

Real World Examples

Simultaneous equations model situations where multiple conditions must be satisfied at once. Here are some practical scenarios:

1. **Business and Economics**: A company sells two types of products, A and B. Product A costs $10, and Product B costs $15. In one day, they sell a total of 50 items and earn $500. How many of each product were sold?

    Let (x) be the number of Product A sold, and (y) be the number of Product B sold.

    Equations:

    $$x + y = 50$$

    $$10x + 15y = 500$$

2. **Ticket Sales**: A theatre sells adult tickets for $12 and child tickets for $8. On one night, 200 tickets are sold, and the revenue is $2,000. How many adult and child tickets were sold?

    Let (a) be the number of adult tickets, and (c) be the number of child tickets.

    Equations:

    $$a + c = 200$$

    $$12a + 8c = 2000$$

3. A coffee shop sells two types of coffee, coffee A and coffee B. Coffee A costs $3 and coffee B costs $4. If the shop sells 300 coffees one day and collects $1000, how many of each type does the shop sell?

   Let (a) be the number of coffee A's sold, and b be the number of coffee B's sold.

   Equations:

   $$a + b = 300$$
   $$3a + 4b = 1000$$

These examples show how simultaneous equations help solve problems with multiple unknowns in real-world contexts. In all of the above equations, real world constraints are expressed as two separate equations, known as simultaneous equations. Solving these equations will help these companies with the exact information they need.

Solving Simultaneous Equations – Substitution Method

The substitution method involves solving one equation for one variable and substituting that expression into the other equation. This reduces the system to a single equation with one variable.

Steps:

1. Express one variable in one of the equations in terms of the other (e.g., express (y) in terms of (x)).

2. Substitute this expression into the other equation.

3. Solve the resulting equation for the remaining variable.

4. Substitute back to find the other variable.

5. Verify the solution in both original equations.

Example:

Solve:

$2x + y = 8$

$x - y = 1$

Step 1: Express the second equation in terms of x:

Add y to both sides, we get:

$x - y + y = 1 + y$

$x = y + 1$

Step 2: Substitute $(x = y + 1)$ into the first equation:

$2(y + 1) + y = 8$

$2y + 2 + y = 8$

Step 3: Simplify and solve for (y):

3y + 2 = 8

Subtract 2 from both sides

3y – 2 = 8 – 2

3y = 6

Divide both sides by 3

y = 2

Step 4: Substitute (y = 2) into (x = y + 1):

x = 2 + 1

x = 3

Step 5: Verify:

First equation: 2(3) + 2 = 6 + 2 = 8 (True)

Second equation: 3 - 2 = 1 (True)

Solution: (x = 3), (y = 2).

Worksheet 16

Solve the following systems using the substitution method. Show all steps and verify your solutions.

1. $x + y = 7$

    $2x - y = 5$

2. $3x + 2y = 11$

    $x - y = 2$

3. $4x - 3y = 1$

   $2x + y = 3$

4. $5x + y = 10$

   $x - 2y = 2$

5. A store sells pens for $2 and notebooks for $5. A customer buys a total of 7 items and spends $26. How many pens and notebooks were bought? Write the system of equations and solve using substitution.

Solving Simultaneous Equation – Elimination Method

The elimination method involves adding or subtracting equations to eliminate one variable, allowing you to solve for the other. This method is often efficient when one of the coefficients is the same in both equations.

Steps:

1. Adjust the equations (if needed) by multiplying so that the coefficients of one variable are equal or opposites.

2. Add or subtract the equations to eliminate one variable.

3. Solve the resulting equation for the remaining variable.

4. Substitute back to find the other variable.

5. Verify the solution in both original equations.

Example:

Solve:

$3x + 2y = 11$…………………………………………………………………..a

$x - y = 2$…………………………………………………………………………b

The two equations are labelled a and b.

Step 1: The coefficients of x and y are different for the two equations. So, we multiply one of the equations by a number to make the coefficients the same. The coefficients of y are '2' in Equation a, and '-1' in Equation b. The coefficients of x are '3' in Equation a, and '1' in Equation b.

To make the x coefficients the same, we need to multiply Equation b by 3.

To make the y coefficients the same, we need to multiply Equation b by -2.

We decide to multiply Equation b by -2, as it's easier to work with a smaller number. After multiplying Equation b by -2, we get the two equations now as:

$3x + 2y = 11$...a

$-2x + 2y = -4$...b

Step 2: Now, let's subtract Equation b from Equation a to eliminate the y-variable.

Equation a – Equation b

$3x + 2y - (-2x + 2y) = 11 - (-4)$

$3x + 2y + 2x - 2y = 11 + 4$

$5x = 15$

Step 3: Now we solve the Equation as we do for any other equation:

$5x = 15$

Dividing both sides by 5, we get:

$x = 3$

Step 4: Now substitute x into 3 into one of the equations to find y.

$x = 3$ in Equation a,

$3(3) + 2y = 11$

$9 + 2y = 11$

Subtracting 9 from both sides

$9 + 2y - 9 = 11 - 9$

$2y = 2$

Dividing both sides by 2,

$y = 1$

So, the answer is (x=3; y=1) or (3,1)

Worksheet 17

Solve the following systems using the elimination method. Show all steps and verify your solutions.

1. $2x + 3y = 8$

    $2x - y = 4$

2. $5x + 2y = 12$

   $3x + 2y = 8$

3. $4x - y = 7$

   $2x + 3y = 7$

4. $6x + 4y = 14$

   $3x - y = 4$

5. A coffee shop sells small coffees for $3 and large coffees for $5. On one day, 50 coffees are sold, and the revenue is $190. How many small and large coffees were sold? Write the system of equations and solve using elimination.

Solving Simultaneous Equations – Graphical Method

The graphical method involves plotting both equations as lines on a coordinate plane. The point where the lines intersect is the solution to the system.

Steps:

1. Rewrite each equation in slope-intercept form ($y = mx + b$).

2. Plot the lines on a graph using the slope (m) and y-intercept (b) as done in the Chapter "Graphing Linear Equations".

3. Identify the point where they intersect (x, y).

4. Verify the solution by substituting into both original equations.

Example:

Solve:

$2x + y = 5$

$x - y = 1$

Step 1: Rewrite in slope-intercept form: $y = mx + c$

First equation: $y = -2x + 5$

Second equation: $y = x - 1$

Step 2: Plot the lines as described in the chapter "Graphing Linear Equations":

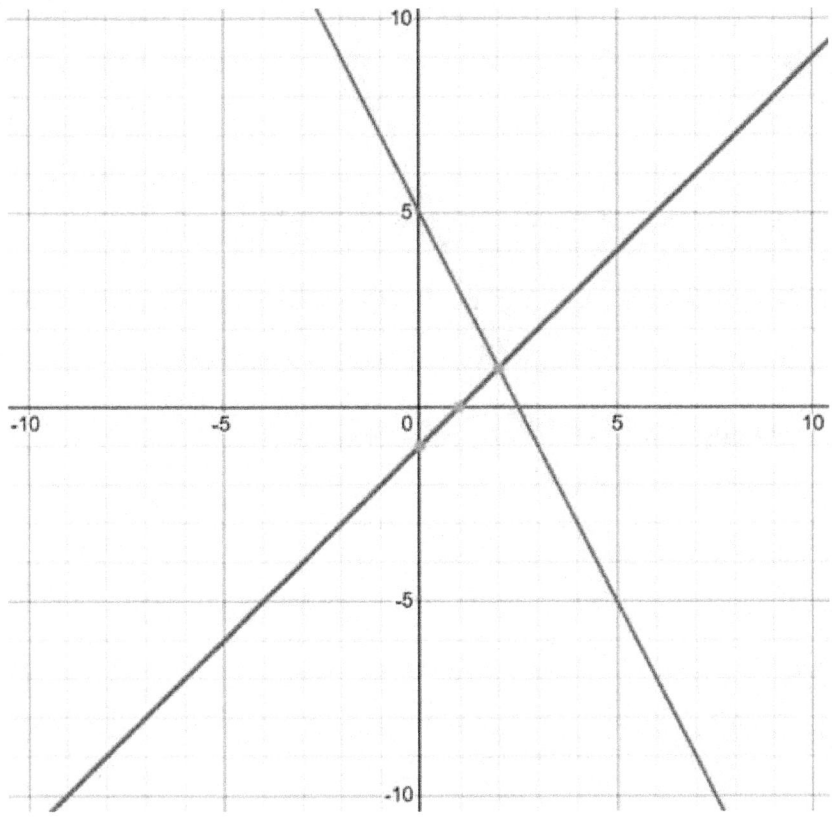

The red line above is y = -2x + 5 and the blue line is y = x - 1

Step 3: Identify the point where the two lines intersect.

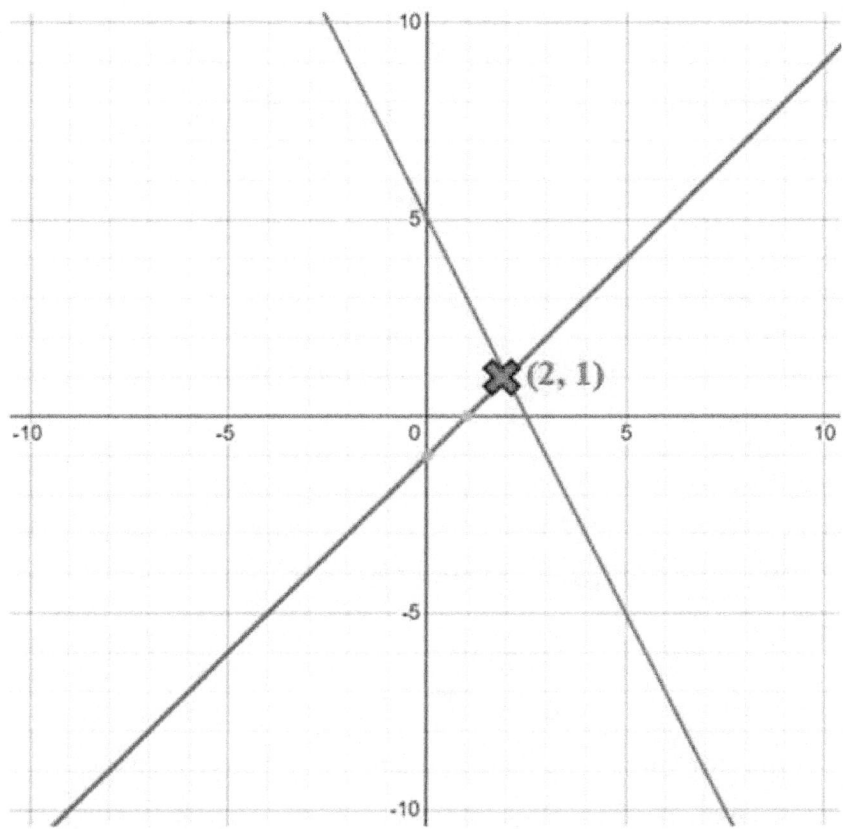

The lines intersect at (2, 1) as shown below.

Step 4: Verify by substituting x = 2; y = 1 into each of the equations:

First equation: (2(2) + 1 = 5) (True)

Second equation: (2 - 1 = 1) (True)

Solution: $(x = 2)$, $(y = 1)$.

Notes:

- If the lines are parallel, there is no solution (inconsistent system).

- If the lines coincide, there are infinitely many solutions (dependent system).

- Graphing is less precise than algebraic methods for non-integer solutions.

Tip: Which method should I use?

Always go with Algebraic Methods, unless asked to do otherwise. It is more accurate and requires less space. I prefer the Elimination method as it has fewer steps, but the substitution method can work as well.

Worksheet 18

Solve the following systems using the graphical method. Sketch the lines on the graphs below and estimate the intersection point, then verify algebraically.

1. $x + y = 4$

    $x - y = 2$

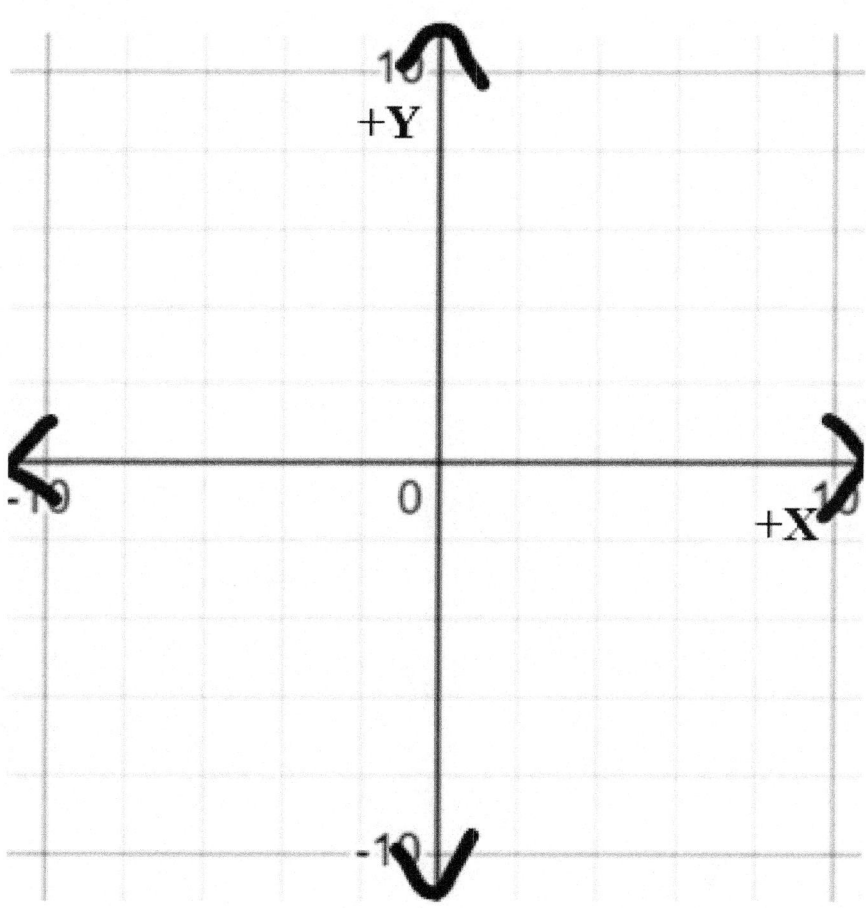

2. $2x + y = 6$

   $x + y = 3$

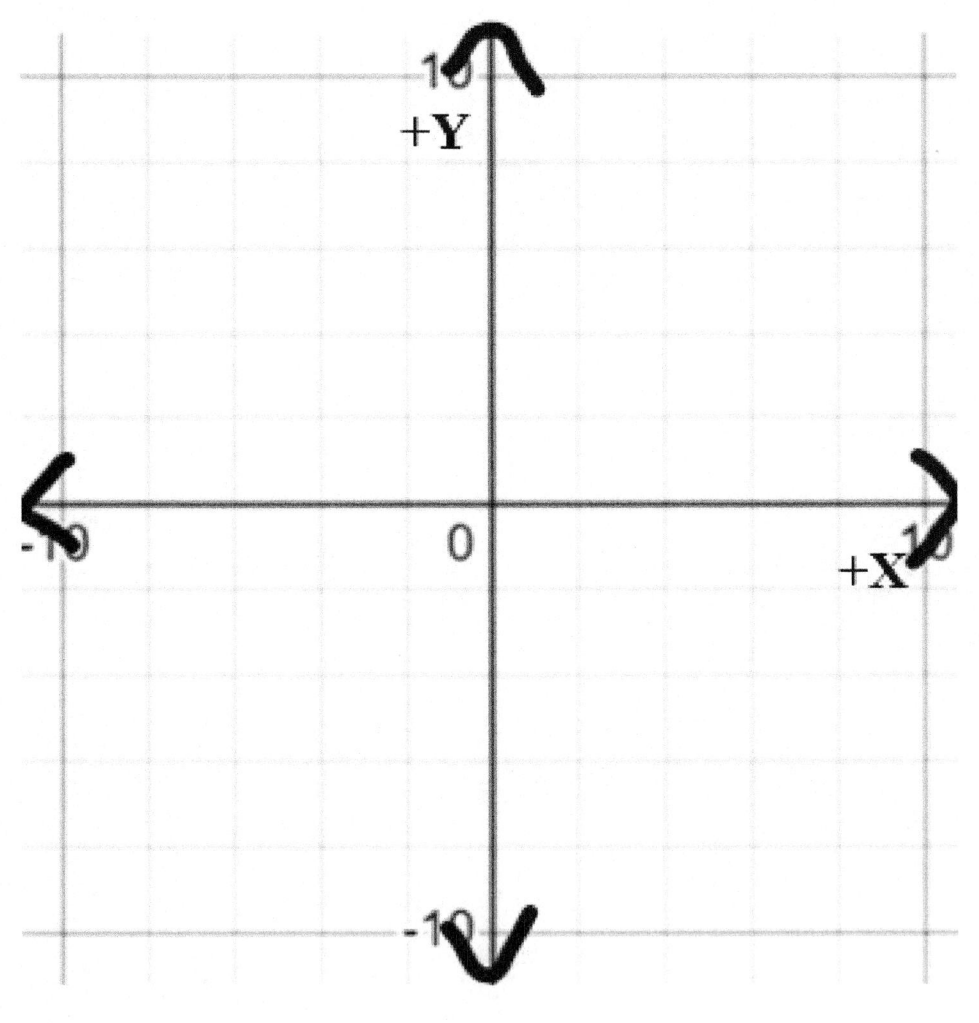

3. $3x - y = 5$

   $x + y = 3$

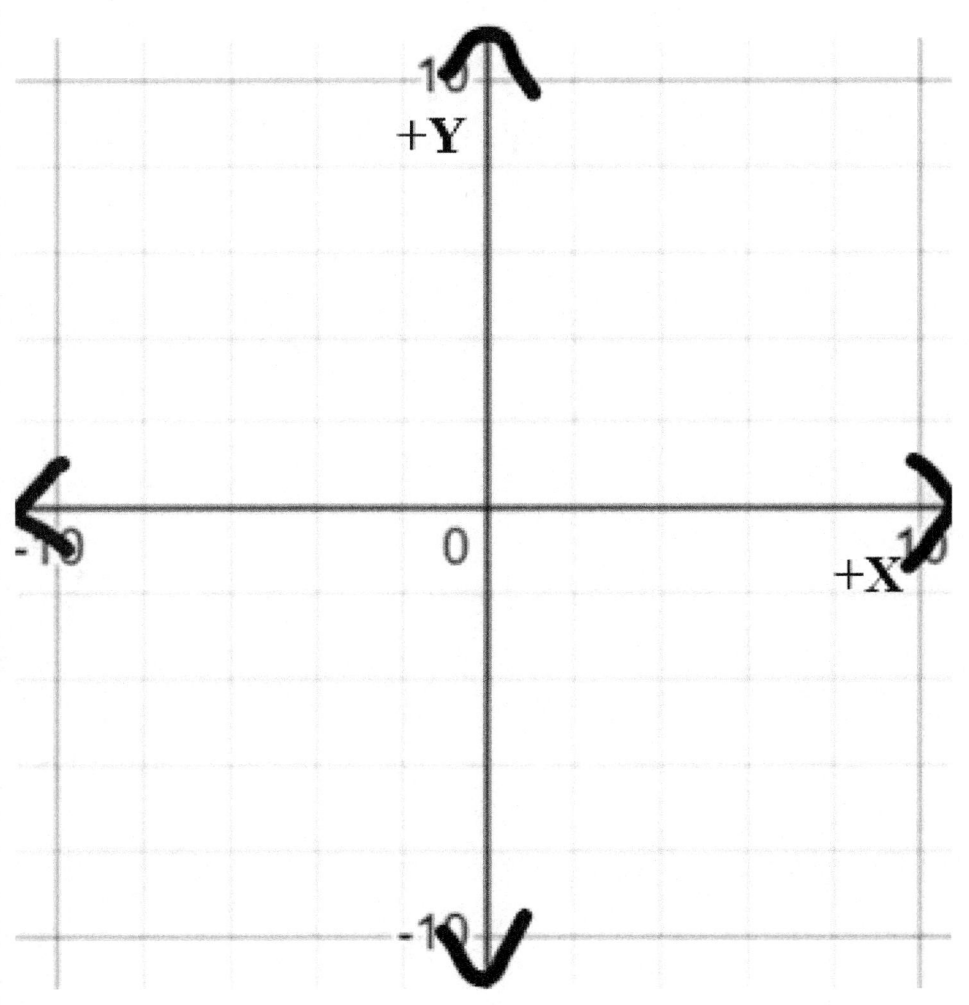

4. $4x + 2y = 8$

$2x - y = 1$

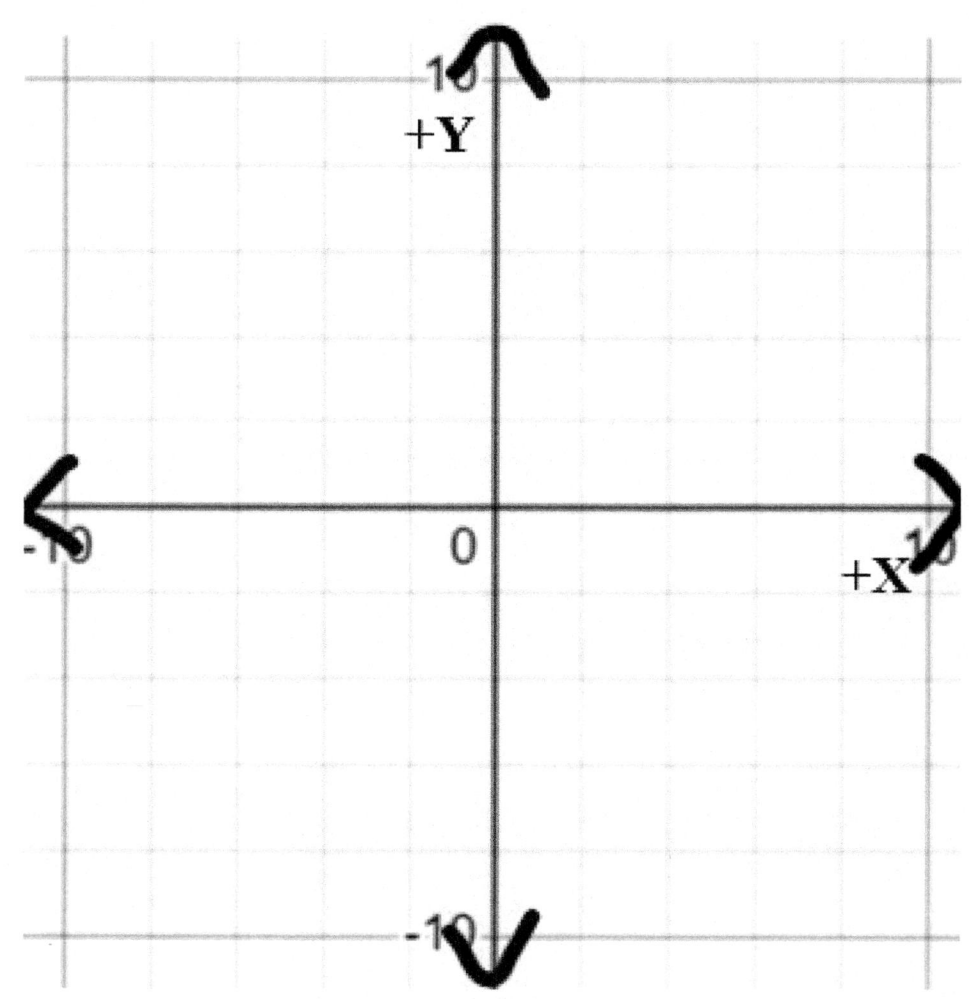

5. A bakery sells muffins for $2 and cookies for $1. On one day, 10 items are sold, and the revenue is $15. How many muffins and cookies were sold? Write the system, solve graphically, and verify.

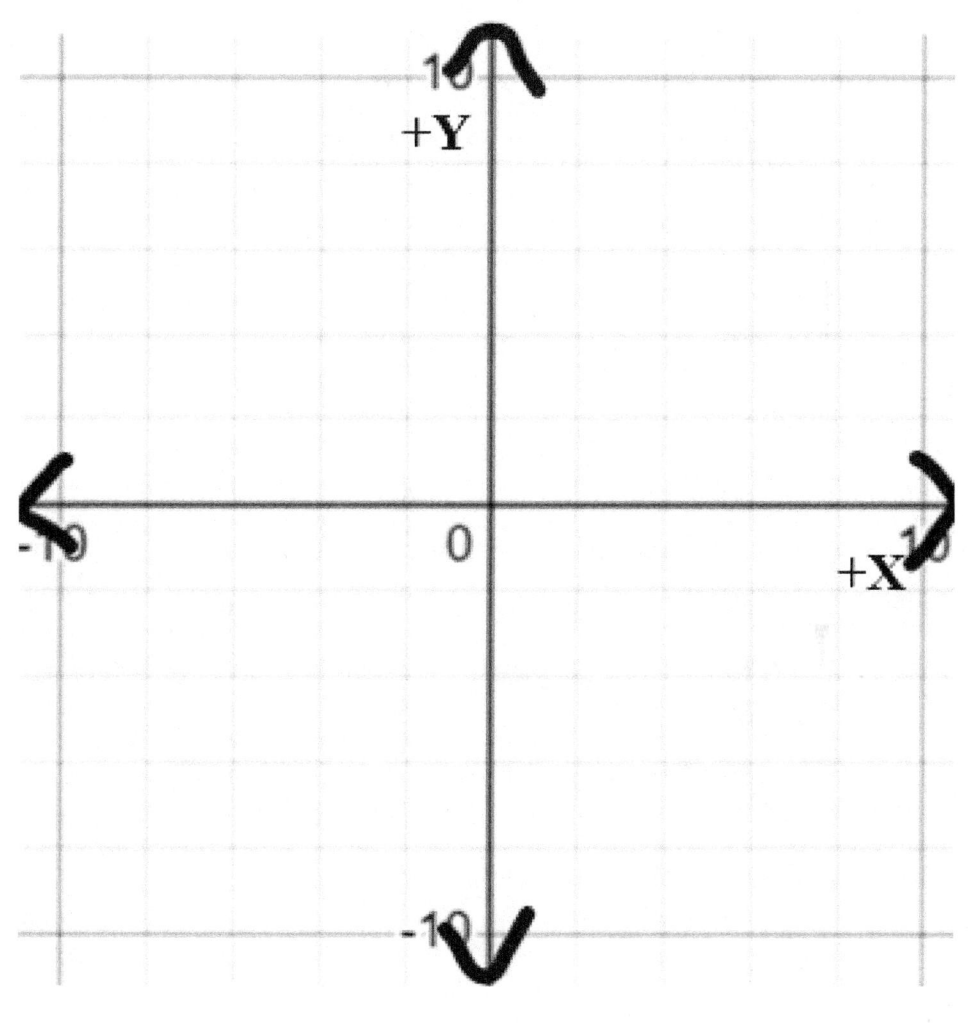

Factorization

What is Factorization?

Factorization is the process of breaking down a mathematical expression, such as a number or a polynomial, into a product of simpler expressions, called factors. These factors, when multiplied together, give the original expression. Think of it like dividing a big puzzle into its individual pieces.

For example:

- The number 12 can be factored as (2×6) or (3×4) or $(2 \times 3 \times 3)$.

- The polynomial $(x^2 + 5x + 6)$ can be factored into $(x + 2)(x + 3)$.

In algebra, factorization is a key skill used to simplify expressions, solve equations, and understand the structure of polynomials. Factorization helps in solving equations, finding common denominators, and simplifying complex expressions, making it a fundamental tool in algebra.

Real World Examples

Factorization is not just a classroom exercise; it has practical applications in everyday life and various fields:

1. A baker splitting 24 cookies equally into 4 boxes is using factorization ($24 = 4 \times 6$).

2. A carpenter needs to make a panel of length 40 inches. He uses 5 smaller wooden pieces (length 8) to make a panel of length 40 inches. This is applying factorization. ($5 \times 8 = 40$).

3. **Finance and Budgeting**: Suppose you're planning a party and need to pack 24 cupcakes and 36 cookies into boxes. Factorizing these numbers:

   $24 = 2 \times 2 \times 2 \times 3 = \mathbf{12} \times 2$

   $36 = 2 \times 2 \times 3 \times 3 = \mathbf{12} \times 3$

   We find the greatest common factor (12) to determine the largest size of box you can use to prevent leftovers.

4. **Computer Science**: Factorization is used in cryptography, such as RSA encryption, where large numbers are factored to secure online transactions. It also helps optimize algorithms by breaking down complex problems into simpler parts.

5. **Engineering**: When designing structures, engineers factor equations to simplify calculations for load distribution, ensuring bridges or buildings are safe.

6. **Time Management**: If you have 60 minutes to complete tasks that take 12 minutes each, factorizing 60 and 12 helps determine how many tasks (5) you can complete.

$60 = 2 \times 2 \times 3 \times \mathbf{5}$

$12 = 2 \times 2 \times 3$

GCF (Greatest Common Factor) Factorization

The Greatest Common Factor (GCF) of two or more numbers or expressions is the largest factor they share. GCF factorization involves factoring out the GCF from each term in an expression to simplify it.

Steps:

1. Identify the GCF of the coefficients and variables in the expression.

2. Divide each term by the GCF.

3. Write the expression as the GCF multiplied by the resulting terms.

Example: Factorize: $12x^2 + 18x$.

Step 1: We factorize the numerical part, and the algebraic part separately and then combine them together.

Coefficients: 12 and 18.

$12 = 2 \times 2 \times 3 = \mathbf{6} \times 2$

$18 = 2 \times 3 \times 3 = \mathbf{6} \times 3$

GCF = 6.

Variables:

$x^2 = \mathbf{x} \times \mathbf{x}$

$x = \mathbf{x} \times 1$

GCF = x.

The GCF of the numerical part is 6, and the algebraic part is x. Combining them, we get a GCF of **6x**

Overall GCF = **6x**.

Step 2: Now we divide each term by the GCF

Divide:

$12x^2$ by 6x = **2x**

$$\frac{12x^2}{6x} = \frac{12x.\cancel{x}}{6\cancel{x}} = 2x$$

18x by 6x = **3.**

$$\frac{18\cancel{x}}{6\cancel{x}} = 3$$

Step 3: Now, we write out the new expression with the GCF multiplied by the resulting terms:

$12x^2 + 18x = \mathbf{6x\ (2x + 3)}$

Why it matters: GCF factorization simplifies expressions and is often the first step in solving equations or factoring more complex polynomials.

Group Factorization

Group factorization involves grouping terms in a polynomial to factor it. This is often used when the polynomial has four or more terms and no single GCF for all terms. You pair terms with common factors and factor each group, then factor out any common binomial.

Steps:

1. Group terms (usually in pairs) that share a common factor.

2. Factor out the GCF from each group and rewrite the expression.

3. If the resulting binomials have a common factor, factor that out.

Example: Factorize: ax + ay + bx + by.

Step 1: We see that the first two terms have a common factor of a, and the last two terms have a common factor of b.

Groups: (ax + ay) + (bx + by).

Step 2: Factor each group:

ax + ay = a (x + y)

$bx + by = b(x + y)$.

$ax + ay + bx + by = a(x + y) + b(x + y)$

Step 3:

Now, let's look at a $(x + y)$ and b $(x + y)$. They have $(x + y)$ in common.

Factor out the common binomial:

$a(x + y) + b(x + y) = (x + y)(a + b)$.

Why it matters: Group factorization is useful for polynomials that don't factor easily with other methods and helps in solving equations.

Worksheet 19

1. Factor each expression using the GCF:

    a. $8x + 12 = $ _____

    b. $15y^2 - 25y = $ _____

    c. $6a^3 + 9a^2 = $ _____

    d. $14m^2n + 21mn^2 = $ _____

    e. $ab^2 + a^2b + abc = $ _____

2. A store has 60 apples and 36 oranges to pack into baskets. What is the largest number of fruits in each basket they can make, if each basket has the same number of apples and oranges, with no fruit left over?

3. Factorize: $24x^3y^2 - 36x^2y + 12xy^2$.

4. Factorize each expression by grouping:

   a. $2x + 2y + ax + ay$

   b. $3m - 3n + km - kn$

c. $x^2 + 2x + xy + 2y$

d. $6a - 3b + 2ab - b^2$

e. $x^2 - xy + 2x - 2y$

5. Factorize: $4x^3 + 2x^2 + 2xy + y$.

6. The area of a rectangular garden is given by the formula $2x^2 + 2xy + x + y$. Factorize the expression to find possible dimensions of the garden.

Quadratic Factorization

Quadratic factorization applies to quadratic expressions. What exactly is a quadratic equation? A quadratic equation is an algebraic expression where the highest order of the variable is 2. For example, the below expression is a quadratic equation for unknown variable x.

$ax^2 + bx + c.$

Now, solving the quadratic expression is slightly different from what we've learnt so far. It involves breaking down the middle term bx into 2 x terms. Once that happens, we have 4 terms and can use group factorization to solve the expression

Let's look at the steps.

Steps:

1. Write $ax^2 + bx + c$.

2. Find the product of two terms a and c. We get the product ac.

3. Find two numbers that multiply to ac and add to b.

4. Let's say these two factors are b_1 and b_2. These replace the middle term bx and we get:

   $$ax^2 + b_1x + b_2x + c = 0$$

5. Now, we solve the above equation using grouping.

Sounds confusing? Don't worry. It'll get easier with an example.

Example:

Factorize $x^2 + 5x + 6$.

Step 1:

This is a quadratic equation of the form: $ax^2 + bx + c$

Here a = 1; b = 5; c = 6

Step 2:

Find the product of a and c: We get ac = 6 × 1 = 6

Step 3:

Now we need to find two numbers that add to 5 (b = 5) and multiply to 6 (ac = 6).

I like to look at the factors of 6; and see what they add to:

$6 \times 1 = 6$; Factors add to 7 (6 + 1) ✗

$3 \times 2 = 6$; Factors add to 5 (3 + 2) ✔

$-6 \times -1 = 6$; Factors add to -7 (-6 - 1) ✗

$-3 \times -2 = 6$; Factors add to -5 (-3 -2) ✗

3 and 2 are the factors of 6 that add to 5

Step 4:

So we replace 5x with (2x + 3x); so our equation becomes

$x^2 + 2x + 3x + 6$

Step 5:

Now we solve the equation using group factorization. We group the first two and last two into separate groups.

$(x^2 + 2x) + (3x + 6)$

In the 1st group, x is common.

$(x^2 + 2x) = x (x + 2)$

In the 2nd group, 3 is common.

$(3x + 6) = 3 (x + 2)$

Now, we rewrite the equation:

$x (x + 2) + 3 (x + 2)$

We can take $(x + 2)$ common now. We get:

$(x + 2) (x + 3)$

$x^2 + 5x + 6 = (x + 2) (x + 3)$

Since this is a new concept, let's look at another example.

Example:

Factorize: $2x^2 + 7x + 3$.

Step 1:

This is a quadratic equation of the form: $ax^2 + bx + c$

Here $a = 2$; $b = 7$; $c = 3$

Step 2:

Find the product of a and c: We get $2 \times 3 = 6$

Step 3:

Now we need to find two numbers that add to 7 (b = 7) and multiply to 6 (ac = 6).

I like to look at the factors of 6; and see what they add to:

$6 \times 1 = 6$; Factors add to 7 (6 + 1) ✔

$3 \times 2 = 6$; Factors add to 5 (3 + 2) ✘

$-6 \times -1 = 6$; Factors add to -7 (-6 - 1) ✘

$-3 \times -2 = 6$; Factors add to -5 (-3 -2) ✘

6 and 1 are the factors of 6 that add to 7

Step 4:

So, we replace 7x with (6x + x); so our equation becomes

$2x^2 + 6x + x + 3$

Step 5:

Now we solve the equation using group factorization. We group the first two and last two into separate groups.

$(2x^2 + 6x) + (x + 3)$

In the 1st group, 2x is common.

$(2x^2 + 6x) = 2x (x + 3)$

In the 2nd group, no element is common, so we take 1 as the common factor.

$(x + 3) = 1 (x + 3)$

Now, we rewrite the equation:

$2x (x + 3) + 1 (x + 3)$

We can take $(x + 3)$ common now. We get:

$(x + 3) (2x + 1)$

So, we get:

$2x^2 + 7x + 3 = (x + 3) (2x + 1)$

Worksheet 20: Quadratic Factorization

Factor each quadratic:

1. $x^2 + 7x + 12$

2. $x^2 - 8x + 15$

3. $3x^2 + 11x + 6$

4. $2x^2 - 5x - 3$

5. $x^2 + x - 30$

6. $x^2 + 5x + 6$

7. $x^2 - 6x + 8$

8. $2x^2 - 5x - 3$

9. $3x^2 - 10x + 8$

10. $x^2 + 3x - 10$

Difference of Squares Factorization

A difference of squares is an algebraic expression where the squares of two algebraic terms or variables are subtracted from each other. It is of the form $a^2 - b^2$. We can easily solve this by using the equation below.

$a^2 - b^2 = (a + b)(a - b)$.

Example:

Factorize:

$x^2 - 16$.

Recognize: $x^2 - 16 = x^2 - 4^2$.

The equation is in the form of $a^2 - b^2$ where a is x and b is 4.

So, we get the factors of $(a + b)$ and $(a - b)$ or $(x + 4)$ and $(x - 4)$

$x^2 - 16 = (x + 4)(x - 4)$.

Example with coefficients:

Factorize:

$9x^2 - 25$.

Recognize: $9x^2 - 25 = (3x)^2 - (5)^2$.

The equation is in the form of $a^2 - b^2$ where a is 3x and b is 5.

So, we get the factors of $(a + b)$ and $(a - b)$ or $(3x + 5)$ and $(3x - 5)$

$9x^2 - 25 = (3x + 5)(3x - 5)$.

Why it matters: This pattern is common in algebra and simplifies expressions quickly.

Perfect Square Factorization

A perfect square is a algebraic expression that is a product of the same algebraic expression. It is in the form of $a^2 + 2ab + b^2$ or $a^2 - 2ab + b^2$.

We can easily solve this by using the equations below.

$a^2 + 2ab + b^2 = (a + b)^2$

$a^2 - 2ab + b^2 = (a - b)^2$.

Example:

Factorize: $x^2 + 6x + 9$.

Recognize

$$x^2 + 6x + 9 = x^2 + 2.3.x + 3^2$$

This is of the form: $a^2 + 2.a.b + b^2$

Where a = 1; b = 3

So, we get $(a + b)^2$ or $(x + 3)^2$

$$x^2 + 6x + 9 = (x + 3)^2$$

Example

Factorize: $4x^2 - 12x + 9$.

Recognize

$$4x^2 - 12x + 9 = (2x)^2 - 2.2x.3 + 3^2$$

This is of the form: $a^2 - 2.a.b + b^2$

Where a = 2; b = 3

So, we get $(a - b)^2$ or $(2x - 3)^2$

$$4x^2 - 12x + 9 = (2x - 3)^2$$

Why it matters:

Recognizing perfect squares simplifies factoring and solving equations.

It is better to memorize these equations for speed, and over time you will recognize them in quadratic polynomials.

Cubic Factorization

A cubic expression is an algebraic expression that is a sum of two cubic variables or expressions. There are of the form $a^3 + b^3$ and $a^3 - b^3$ and are represented by below equations.

$a^3 + b^3 = (a + b)(a^2 - ab + b^2)$

$a^3 - b^3 = (a - b)(a^2 + ab + b^2)$.

Example: Factor $(x^3 + 8)$.

Recognize: $x^3 + 8 = x^3 + 2^3$.

It is in the form of the equation $a^3 + b^3$ where a is x and b is 2. So, we get:

$(x^3 + 8) = (x + 2)(x^2 - 2x + 4)$.

Example: Factor $(27x^3 - 1)$.

Recognize $27x^3 - 1 = 3x^3 - 1^3$.

It is in the form of the equation $a^3 - b^3$ where a is 3x and b is 1. So, we get:

$27x^3 - 1 = (3x - 1)(9x^2 + 3x + 1)$.

Why it matters: Perfect cube factorization is useful in advanced algebra and calculus for simplifying expressions and solving cubic equations.

Worksheet 21

Factorize the below expressions:

1. $x^2 + 6x + 9$

2. $4x^2 + 12x + 9$

3. $x^2 - 10x + 25$

4. $9x^2 + 6x + 1$

5. $16x^2 - 24x + 9$

6. $x^2 - 4$

7. $9x^2 - 25$

8. $4x^2 - 49$

9. $x^2 - 100$

10. $25x^2 - 16$

11. $x^3 - 64$

12. $27x^3 + 216$

Solving Quadratic Equations

What is a Quadratic Equation?

As we already discussed in the previous chapter, a quadratic equation is an algebraic equation where highest order of variables in the equation is 2.

It is written in the standard form:

$ax^2 + bx + c = 0$

where:

- a, b, and c are constants, with $a \neq 0$ (if $a = 0$, the equation is linear, not quadratic).

- x is the unknown variable we solve for.

The solutions, or roots, are the values of x that make the equation true. A quadratic equation can have:

- Two distinct real roots. For example, a quadratic equation can have the factors $(x - 2)(x + 3)$ with roots of $x = 2$, $x = -3$.

- One repeated real root. This means that the equation is a perfect square. For example, $x^2 + 2x + 1$ has (x-1) as its only factor but repeated since it is a perfect square. So, x = 1 is the root.

- Two complex roots (involving imaginary numbers). In some equations, we have roots where $x = \sqrt{-3}$, which are not real numbers and cannot be drawn on a graph.

When graphed, a quadratic equation forms a parabola, and the roots are where the parabola crosses the x-axis. This helps visualize the number of solutions. This is shown below for the equation x^2+4x+3 with roots of x = -3 and x = -1. You can see that the graph slopes upwards and crosses the x-axis at the roots of x = -3 and x = -1.

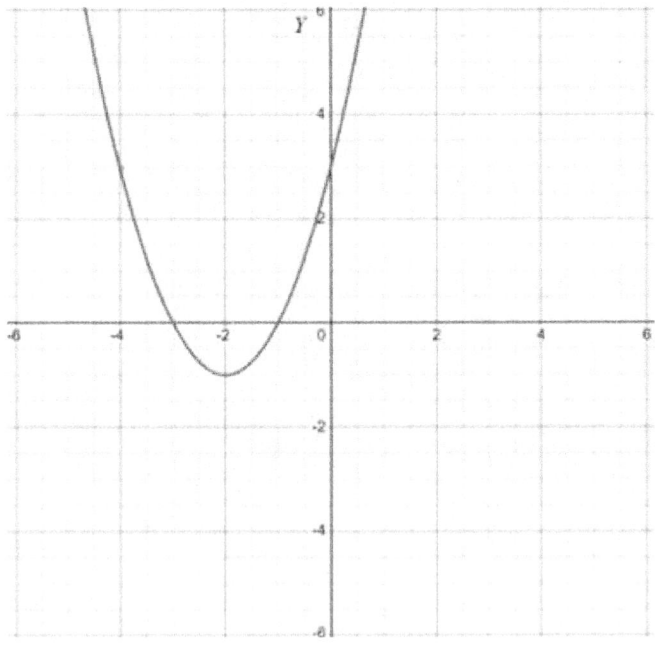

We will discuss these graphs in more detail on the chapter on "Graphing Non-Linear Equations".

Real World Examples

Quadratic equations are used extensively in the real world where it is difficult to model relationships between items with simple linear equations. In fields like physics, motion, optimization, we often need at least a quadratic equation to express the relationship between variables.

Here are a few examples.

Projectile Motion in Physics: The equation of the height that a ball reaches when thrown upwards is given by the equation:

$S = 0.5 \, gt^2 + ut$

Where t is the variable we need to solve for.

S is the distance that the ball travels vertically.

g is the constant known as the acceleration due to gravity and is equal to 9.8 m/s2

u is the initial velocity that the ball is thrown at and is a constant in this equation.

This equation is used in projectile motion and in rocket launches to determine the velocity of the rocket needed to escape the atmosphere.

Area Optimization: In some cases, we need to bend a certain quantity of an item into a rectangular shape. This happens when we're building a fence, for example. The rectangular shape is best modelled using quadratic equations, as we'll see in example below. For example, a gardener has 60 meters of fencing for a rectangular plot. The rectangular plot has an area of 100 m^2.

So, let's say he uses x meters for one side of the fence (the length). So, 2x for the two lengths of the rectangular fence. Then he needs to use (60-2x) for both sides of the breadth. We need to divide that by 2 to get $(60 - 2x)/2$ or $(30 - x)$.

The area is:

A = x (30 - x), where x is the width. Solving for the maximum area uses quadratics.

x(30-x) = 100

-x^2 + 30x = 100

x^2 – 30x + 100 = 0

Solving for x will give us the length of the rectangular fencing.

Profit Analysis: In some cases, a store's profit is modelled by a quadratic equation. Let's say the store's profit from selling x items is given by the equation $P(x) = x^2 - 150x + 2000$. Solving $P(x) = 0$ finds the break-even points, which are the minimum amounts that the store needs to sell to be profitable.

Solving Equations by Factorization

We already went through factorization of quadratics in the previous chapter, but we will do so again. Repetition is one of the best ways to learn.

Let's look at the steps again.

Steps:

1. Write $ax^2 + bx + c$.

2. Frind the product of two terms a and c. We get ac.

3. Find two numbers that multiply to ac and add to b.

4. Let's say these two factors are b_1 and b_2. These replace the middle term bx and we get:

    $ax^2 + b_1x + b_2x + c = 0$

5. Now, we solve the above equation using grouping.

6. In the final step we equate the factors to 0 and find the values of x.

Sounds confusing? Don't worry. It'll get easier with an example.

Example:

Factorize $3x^2 + 8x + 5$.

Step 1:

This is a quadratic equation of the form: $ax^2 + bx + c$

Here a = 3; b = 8; c = 5

Step 2:

Find the product of a and c: We get ac = 5 × 3 = 15

Step 3:

Now we need to find two numbers that add to 8 (b = 8) and multiply to 15 (ac = 15).

I like to look at the factors of 15; and see what they add to:

15 × 1 = 16; Factors add to 16 (15 + 1) ✗

5 × 3 = 15; Factors add to 8 (3 + 5) ✓

-15 × -1 = 15; Factors add to -16 (-15 - 1) ✗

-5 × -3 = 15; Factors add to -8 (-3 - 5) ✗

5 and 3 are the factors of 15 that add to 8

Step 4:

So we replace 8x with (3x + 5x); so our equation becomes

$3x^2 + 3x + 5x + 5$

Step 5:

Now we solve the equation using group factorization. We group the first two and last two into separate groups.

$(3x^2 + 3x) + (5x + 5)$

In the 1st group, 3x is common.

$(3x^2 + 3x) = 3x (x + 1)$

In the 2nd group, 5 is common.

$(5x + 5) = 5 (x + 1)$

Now, we rewrite the equation:

$3x (x + 1) + 5 (x + 1)$

We can take (x + 1) common now. We get:

$(x + 1) (3x + 5)$

$3x^2 + 8x + 5 = (x + 1) (3x + 5)$

Step 6:

We set the factorized equation to 0 to find the values of x.

$(x + 1) (3x + 5) = 0$

So,

$x + 1 = 0$

We get, $x = -1$

Or

$3x + 5 = 0$

$x = -5/3$

So, our roots are **$x = -1$ or $x = -5/3$.**

Now, let's at another example of quadratic equations.

$x^2 - 12x + 20 = 0$

This time we'll let you solve Steps 1 to 5 based on what we've done in previous examples.

After Steps 1 to 5 we get two factors $(x - 10)$ and $(x - 2)$ as the factors.
So,

$(x - 10) (x - 2) = 0$

| $x - 10 = 0$ | or | $x - 2 = 0$ |
| $x = 10$ | or | $x = 2$ |

So, the roots are **x = 10** and **x = 2**.

Worksheet 22

Solve the following equations by factorization. Show all steps and verify solutions by substituting back. Answers are below.

1. $x^2 + 8x + 15 = 0$

2. $x^2 - 7x + 12 = 0$

3. $x^2 - 10x + 25 = 0$

4. $2x^2 + 5x + 3 = 0$

5. $4x^2 - 12x + 8 = 0$

Quadratic Formula

There are many times you would notice that it is impossible or very difficult to factorize the quadratic. What if the quadratic is something like $x^2 + 93x + 850$?

When you compare it to the formula $ax^2 + bx + c = 0$, then $a = 1$ and $c = 850$. $ac = 850$. It would take forever to find all the factors of 850 (there are 12 different ways to multiply numbers to get to 850).

In such cases, we use something called the quadratic formula below:

$$x = \frac{-b \pm \sqrt{(b^2 - 4ac)}}{2a}$$

where:

a, b, c are coefficients from standard form of the quadratic $ax^2 + bx + c$.

$b^2 - 4ac$ is called the discriminant. It determines root types:

If $b^2 - 4ac > 0$, it means there are two distinct real roots.

If $b^2 - 4ac = 0$, it means that there is only one repeated real root. It means the equation is a perfect square.

If $b^2 - 4ac < 0$, it means there are two imaginary roots. It's not possible to plot the solutions on a real number line. **We're not going to solve problems on this type of root as imaginary numbers are not part of**

the algebra beginner syllabus. It's more of an advanced algebra concept.

Now let's look at an example.

Example 1:

Solve: $x^2 + 93x + 850$

$a = 1$, $b = 93$, $c = 850$.

Discriminant: $b^2 - 4ac = (93)^2 - 4*1*850 = 5249$.

Since the discriminant $b^2 - 4ac > 0$, the roots are real. Now let's go ahead and solve this problem.

$$x = \frac{-b \pm \sqrt{(b^2 - 4ac)}}{2a}$$

$$x = \frac{-93 \pm \sqrt{(53^2 - 4 \times 1 \times 850)}}{2 \times 1}$$

$$x = -\frac{-93 \pm 72.45}{2}$$

$$x = -46.5 \pm 36.25$$

So the $\pm$ means that there are two roots when we apply the $+$ and $-$ portions of the equations. So, we have:

$x_1 = -46.5 + 36.23$ $\qquad\qquad$ $x_2 = -46.5 - 36.24$

$x_1 = -10.27$ $x_2 = -82.73$

So, the roots are **x = -10.27** and **x = -82.73**.

Example 2:

Solve $2x^2 + 5x - 3 = 0$.

$a = 2, b = 5, c = -3$.

Discriminant

b^2 - 4ac = 5^2 - 4 $\times$ 2 $\times$ -3 = 25 + 24 = 49.

Since the discriminant b^2 - 4ac > 0, the roots are real. Now let's go ahead and solve this problem.

$$x = \frac{-b \pm \sqrt{(b^2 - 4ac)}}{2a}$$

$$x = \frac{-5 \pm \sqrt{(5^2 - 4 \times 2 \times -3)}}{2 \times 2}$$

$$x = \frac{-5 \pm \sqrt{(49)}}{4}$$

$$x = \frac{-5 \pm 7}{4}$$

$x_1 = 2/4$ $x_2 = -12/4$

$x_1 = 0.5$ $x_2 = -3$

159

So, the roots are x = **0.5** and x = **-3**.

Example 3:

Solve $x^2 + 6x + 9 = 0$.

a = 1, b = 6, c = 9.

Discriminant

$b^2 - 4ac = 6^2 - 4 \times 1 \times 9 = 36 - 36 = 0$.

Since the discriminant $b^2 - 4ac = 0$, the roots are real and repeated. So, it's a perfect square quadratic with one root.

$$x = \frac{-b \pm \sqrt{(b^2 - 4ac)}}{2a}$$

$$x = \frac{-6 \pm \sqrt{(6^2 - 4 \times 1 \times 9)}}{2 \times 1} = \frac{-6 \pm \sqrt{(0)}}{2 \times 1}$$

$$x = -3$$

So, the repeated root is x = **-3**.

Tip: Simplify radicals and fractions carefully. Check the discriminant to predict root types.

Hint: Which Method Should You Use?

If you are unsure of which method to use for a particular method, always go for the quadratic formula method. It works for all quadratics, whereas factorization only works when there are clear factors available.

Worksheet 23

Solve the following equations using the quadratic formula. Simplify answers and indicate if roots are real or complex. Answers are below.

1. $x^2 - 2x - 3 = 0$

2. $4x^2 + 8x + 4 = 0$

3. $x^2 + 3x + 1 = 0$

4. $2x^2 - 8x + 3 = 0$

5. $3x^2 - 6x + 3 = 0$

6. $x^2 + x - 1 = 0$

7. $5x^2 - 7x + 2 = 0$

8. $x^2 - 4x + 4 = 0$

9. $2x^2 - 5x - 3 = 0$

10. $x^2 + 5x + 2 = 0$

Graphing Non-Linear Equations

Difference between Linear and Non-Linear Equations

Linear equations form straight lines on a graph, while non-linear equations do not. They form curves, parabolas, hyperbolas, exponentials etc. Non-linear equations more closely resemble the complexities of the real world, such as planetary orbits, projectile motion, population growth, business statistics, where rates of change vary widely.

This chapter goes into visualizing and graphing a few different non-linear equations, such as quadratic equations and circle equations. We will mention a few other equations without going into details, as those equations are beyond the scope of this book.

A linear equation in two variables, x and y, has the form $y = mx + c$, where m and c are constants, and variables appear only to the first power. Its graph is a straight line with a constant slope m. For example, $y = 2x + 3$ is a line with slope 2 and y-intercept 3. It has a constant rate of change, like steady speed or linear savings growth.

A non-linear equation involves terms where variables are raised to powers other than 1, multiplied together, or embedded in functions like square roots, exponentials, or trigonometric functions. Examples include $x^2 + y^2 = 16$, or $y = e^x$. Their graphs are curves, reflecting variable rates of change. Non-linear equations model dynamic systems, such as the arc of a thrown ball or viral growth.

Recognizing these differences helps you choose appropriate graphing strategies and interpret the behaviour of mathematical models.

Quadratic Equation Graphs

As mentioned in the previous chapter, a quadratic equation is an equation where the highest order of the variable is 2. The equation is of the form $y = ax^2 + bx + c$, where a, b, c are constant values.

The graph is a parabola, a U-shaped curve or inverted U-shaped curve. Parabolas model scenarios like the trajectory of a soccer ball, satellite dish shapes, or profit functions in business.

Shapes of Quadratic Equations

The first thing we need to understand is that the shape of the quadratic equation varies depending on the value of variable a in the equation $y = ax^2 + bx + c$.

If a < 0, we get an inverted U-shaped parabola.

If a > 0, we get a U-shaped parabola.

For example, the equation $y = x^2 + 5x + 6$ has a = 1. In this case, a > 0, so we get a U-shaped parabola as shown below.

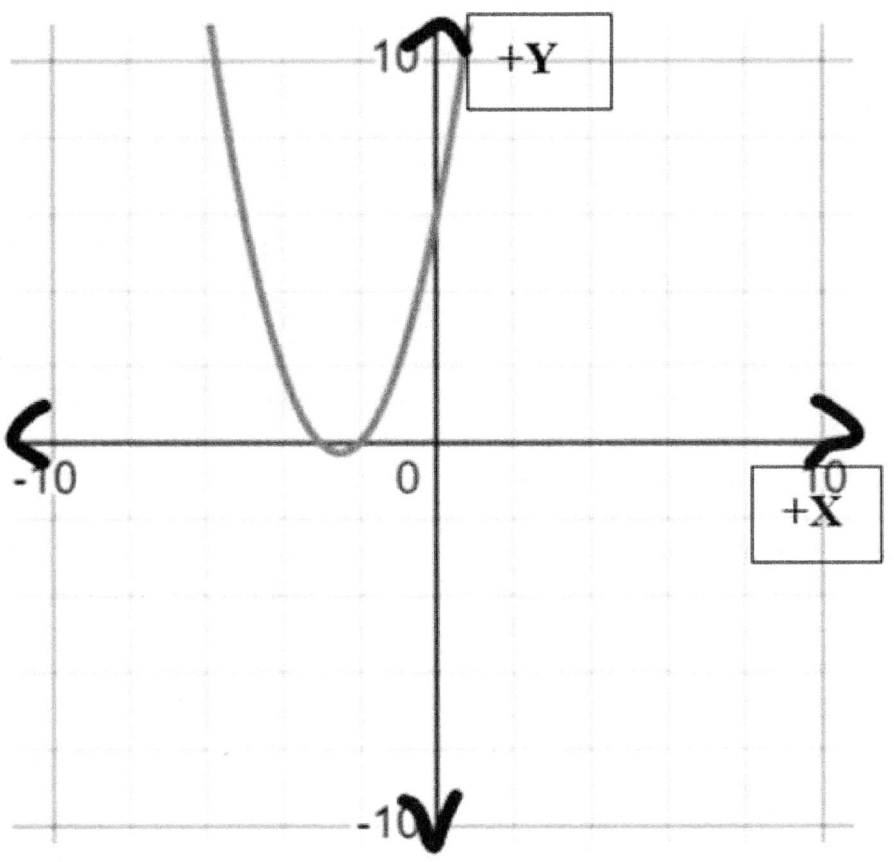

Now, what if we have $y = -x^2 + 5x + 6$. a = -1 which means a < 0, and we have an inverted U-shaped parabola as shown below.

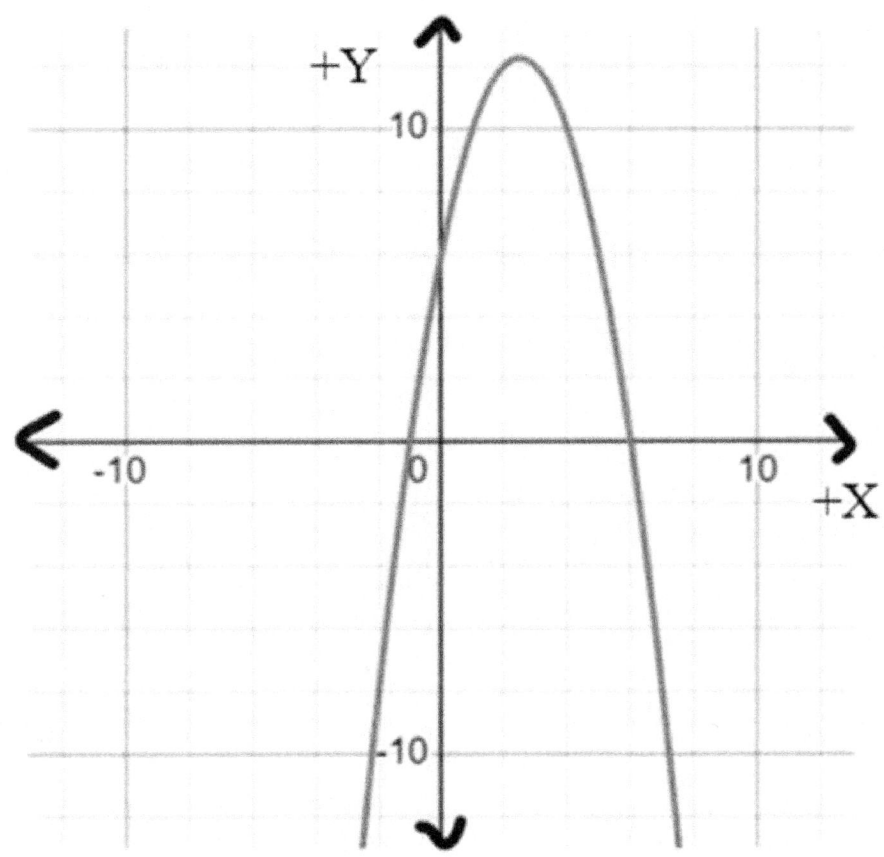

Now, what if we have an equation where $x = ay^2 + bx + c$. It's in the same form as the previous equation but the axes are inverted. The squares are on the y-side, so the parabolas on the graph will be from right/left instead of up/down. There are two forms of parabolas for the equation $x = ay^2 + bx + c$.

If $a < 0$, we get a right curved parabola. Think of it as a U-shaped parabola rotated 90 degrees to the right.

169

If a > 0, we get a left curved parabola. Think of it as a U-shaped parabola rotated 90 degrees to the left.

Now, let's look at the example of the equation x = y² + 5y + 6. It has a>0 so it has a right curved parabola as shown below.

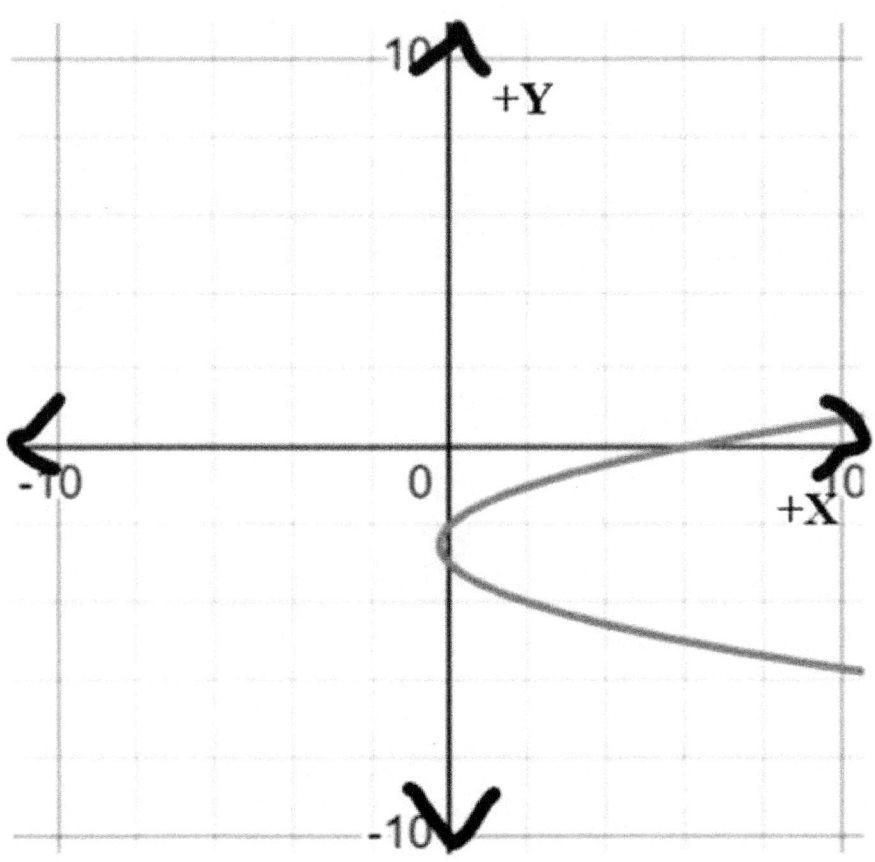

Now, what if x = -y² + 5y + 6. Since a <0, we have a left curved parabola.

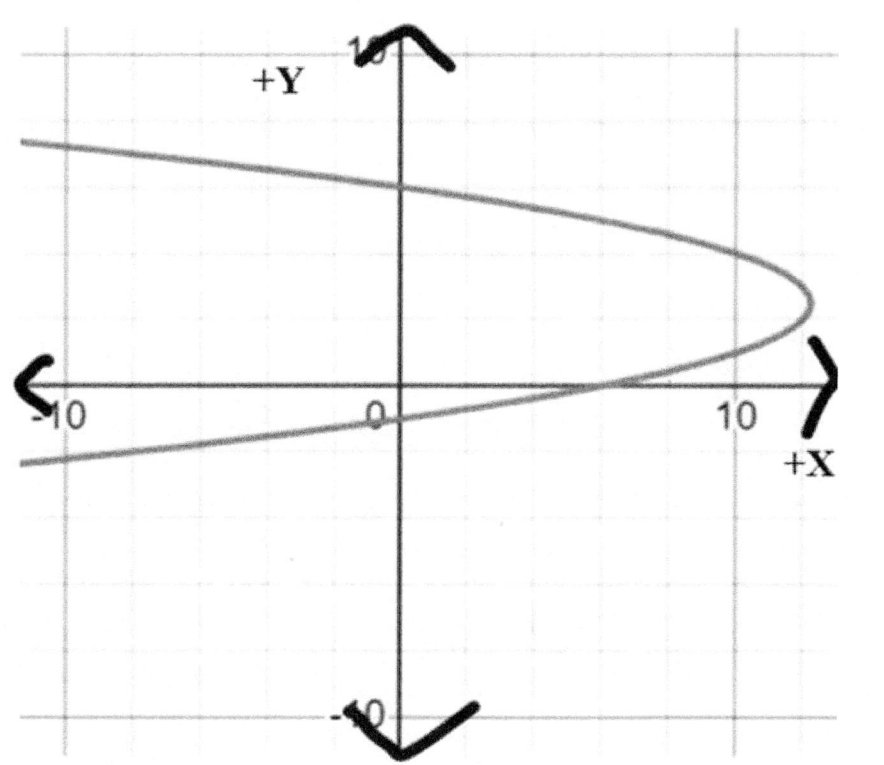

Steps to Graphing a Parabola $y = ax^2 + bx + c$

Now that we understand how quadratic parabolas are shaped based on the equations, let's look at the different steps we need to take to graph a parabola of the standard form $y = ax^2 + bx + c$.

Step 1: Determine the direction of curvature of the parabola. Is it U-shaped, inverted U-shaped, right curved parabola or left curved parabola? Figure this out based on the graphs above in "Shapes of Quadratic Equations".

Step 2: Compute the co-ordinates of the vertex of the parabola. For the-coordinate in the equation $y = ax^2 + bx + c$, we find the x-coordinate using:

$$x = \frac{-b}{2a}$$

Once we get the x-coordinate, substitute it in the equation, then find y.

Step 3: Find the y-intercept of the curve: Set $x = 0$.

Step 4: Find x-intercepts: Solve the quadratic equation $ax^2 + bx + c = 0$.

Step 5: Draw the curve: Join all the points in Steps 2-4, and connect points smoothly and extend the graph, ensuring symmetry about the axis.

Step 6 (Optional): If we want to ensure that the graph is smoother, we can plot additional points. Choose a couple of x-values on both sides of the vertex (e.g., $x = -b/(2a) \pm 1$) and find the y-values.

Now, let's look at an example.

Draw a graph for $y = 2x^2 - 4x - 6$.

In this case, $a = 2$, $b = -4$, $c = -6$

Step 1: Since the equation is of the form $y = ax^2 + bx + c$ and a>0, the equation is a U-shaped parabola.

172

Step 2:

x-coordinate of Vertex $= \dfrac{-b}{2a} = \dfrac{-(-4)}{2(2)} = 1$

We use $x = 1$ to get y-coordinate of vertex

y-coordinate of Vertex $= 2(1)^2 - 4(1) - 6 = -8$

So, the vertex is (1, -8),

Step 3:

For the y-intercept we plug $x = 0$ into the equation.

$y = 2(0)^2 - 4(0) - 6 = -6$

So, the y-intercept is (0, -6)/

Step 4:

For the x-Intercepts, we solve $2x^2 - 4x - 6 = 0$.

Let's solve this equation using the quadratic formula:

$$x = \dfrac{-b \pm \sqrt{(b^2 - 4ac)}}{2a} = \dfrac{-(-4) \pm \sqrt{(-4^2 - 4.2.-6)}}{2.2} = \dfrac{4 \pm 8}{4}$$

We get $x = 3$ and $x = -1$ as the x-intercepts.

So, the x-intercept points are (3,0) and (-1,0).

So, now let's draw the graph below.

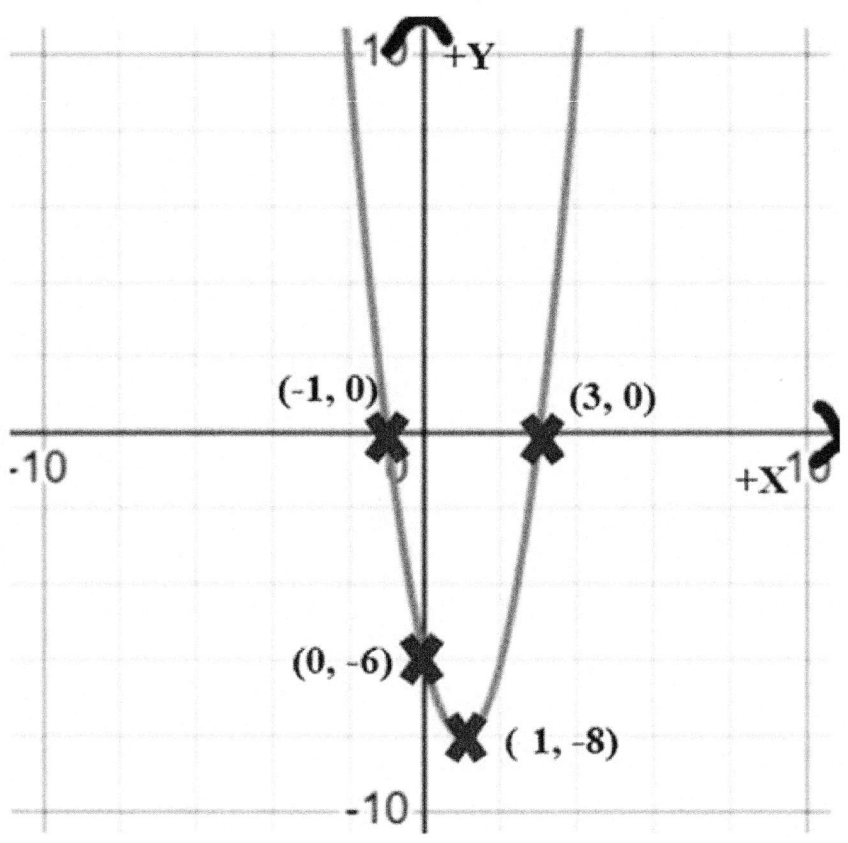

What if we have a different form of the parabola? We have a different set of steps as below:

Steps to Graphing a Parabola x = ay² + by + c

Now that we understand how quadratic parabolas are shaped based on the equations, let's look at the different steps we need to take to graph a parabola of the standard form $x = ay^2 + by + c$.

Step 1: Determine the direction of curvature of the parabola. Is it U-shaped, inverted U-shaped, right curved parabola or left curved parabola? Figure this out based on the graphs above in "Shapes of Quadratic Equations".

Step 2: Compute the co-ordinates of the vertex of the parabola. For the-coordinate in the equation $x = ay^2 + by + c$, we find the y-coordinate using:

$$y = \frac{-b}{2a}$$

Once we get the x-coordinate, substitute it in the equation, then find y.

Step 3: Find the x-intercept: Set y = 0.

Step 4: Find y-intercepts: Solve the quadratic equation $ay^2 + by + c = 0$.

Step 5: Draw the curve: Join all the points in Steps 2-4, and connect points smoothly and extend the graph, ensuring symmetry about the axis.

Step 6 (Optional): If we want to ensure that the graph is smoother, we can plot additional points. Choose a couple of x-values on both sides of the vertex (e.g., $y = -b/(2a) \pm 1$) and find the y-values.

Now, let's look at an example of this:

$x = y^2 + 2y + 1$

In this case, a = 1, b = 2, c = 1

Step 1: Since the equation is of the form $x = ay^2 + by + c$ and a > 0, the equation is a right curved parabola.

Step 2:

y-coordinate of Vertex $= \dfrac{-b}{2a} = \dfrac{-(2)}{2(1)} = -1$

We insert y = -1 to get y-coordinate of vertex

x-coordinate of Vertex $= (-1)^2 + 2(-1) + 1 = 0$

So, the vertex is **(-1, 0),**

Step 3:

For the x-intercept we plug y = 0 into the equation.

$x = (0)^2 + 2(0) + 1 = 1$

So, the x-intercept is **(1, 0).**

Step 4:

For the y-Intercepts, we solve $y^2 + 2y + 1 = 0$.

Let's solve this equation using the quadratic formula:

$$y = \frac{-b \pm \sqrt{(b^2 - 4ac)}}{2a} = \frac{-(2) \pm \sqrt{(2^2 - 4.1.1)}}{2.1} = \frac{-2 \pm 0}{2}$$

We get y = -1 as the single x-intercept. This means that this equation is a perfect square. So, the y-intercept point is **(0, -1)**.

So, now let's draw the graph below.

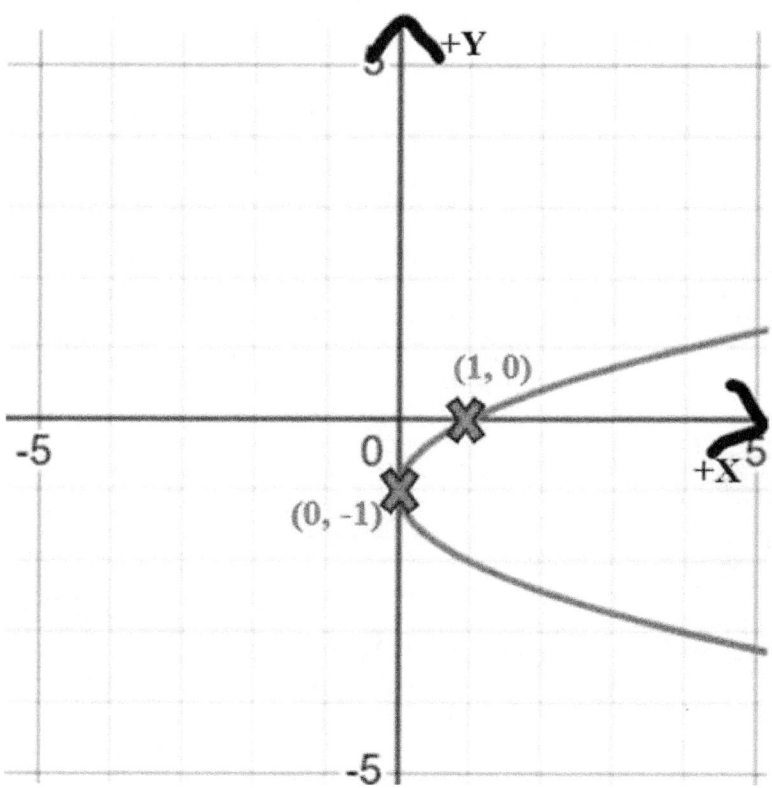

Worksheet 24

Instructions: For each quadratic equation, identify the vertex, axis of symmetry, y-intercept, and x-intercepts (if any). Graph the parabola on a grid. Show all work.

1. $y = x^2 + 2x - 3$

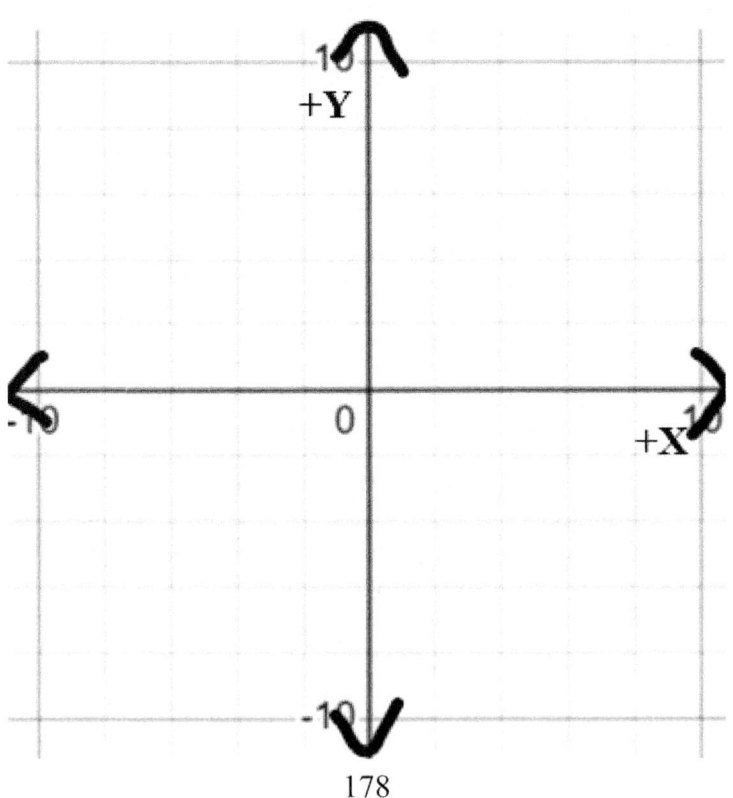

178

2. $y = -2x^2 + 8x - 6$

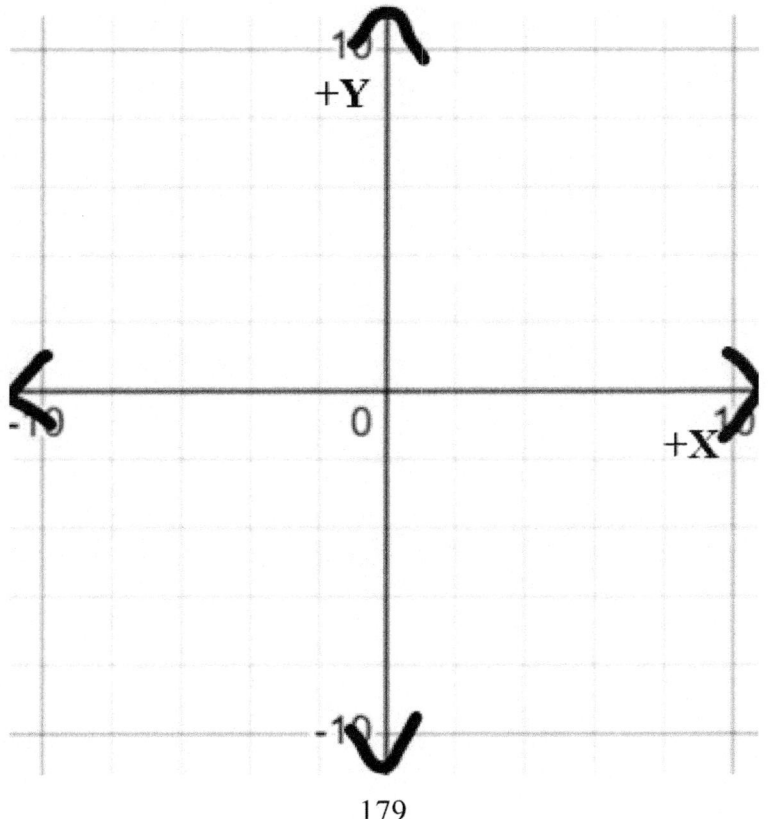

3. $x = 2y^2$

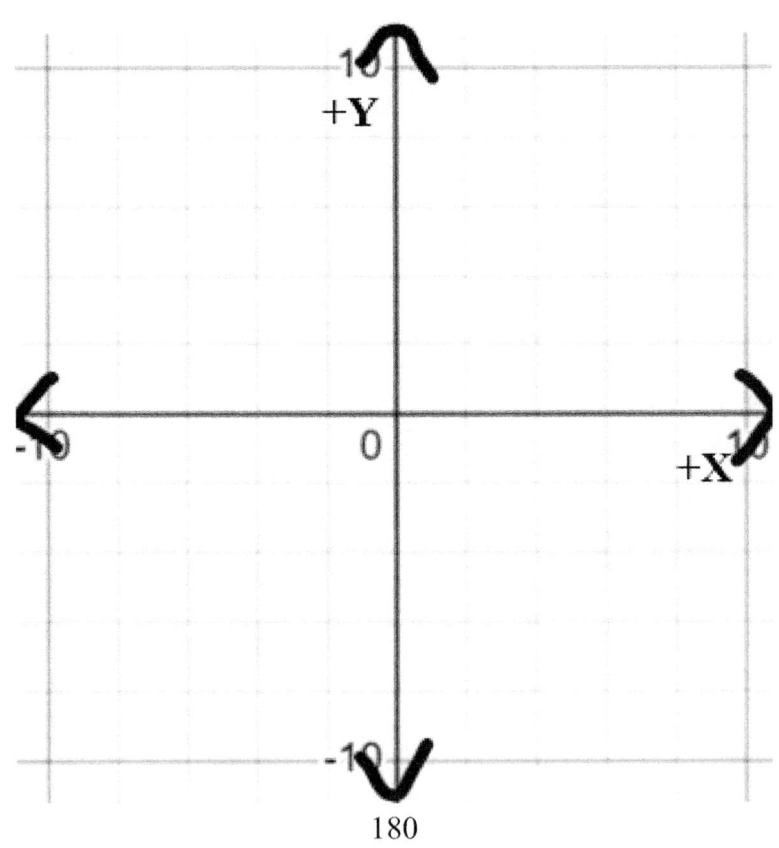

4. $x = y^2 + 4y + 3$

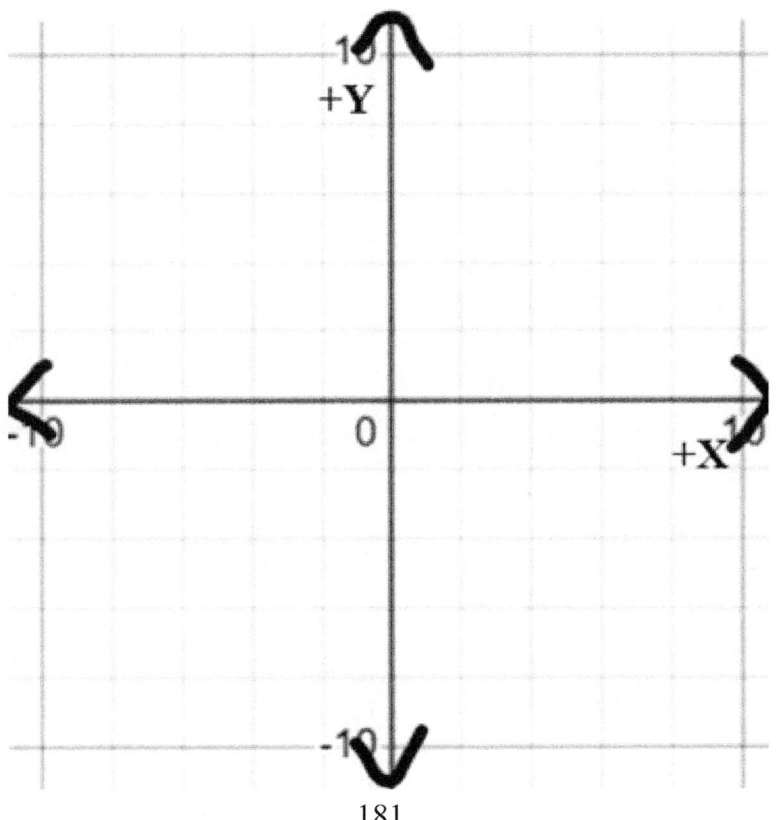

5. The height of a rocket is given by h(t) = -8t² + 32t + 5 (in feet, t in seconds). Find the maximum height and when it hits the ground. Label the axes and sketch the graph.

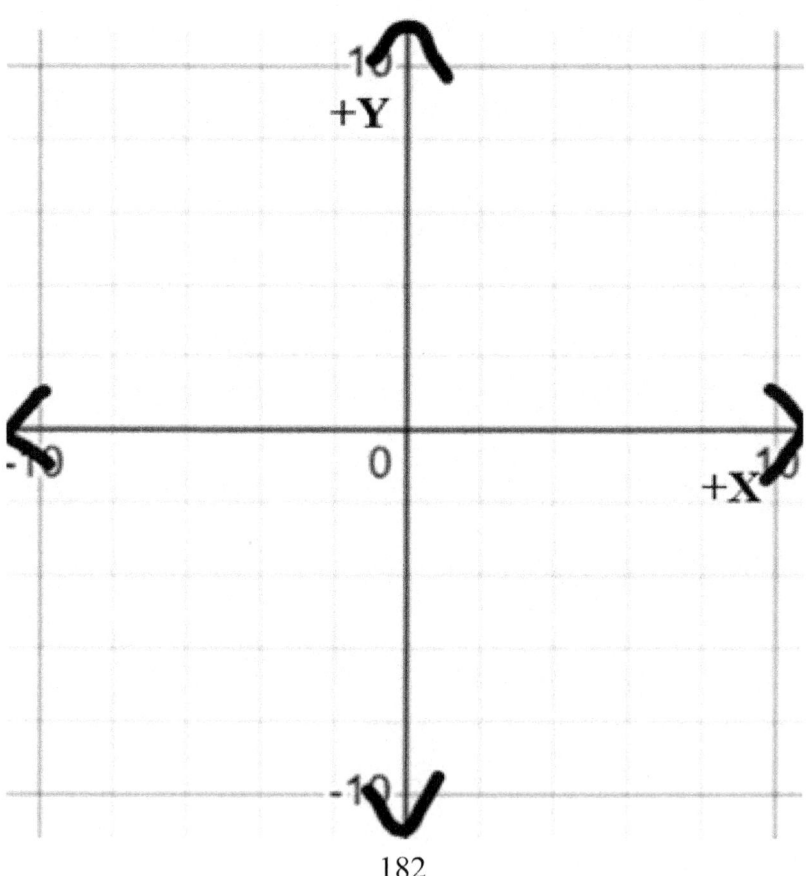

Equation of a Circle

The equation of a circle represents a set of points that are of equal distance from a fixed point. The fixed point is the center of the circle, and the distance is the radius.

The standard form equation for a circle is:

$(x - h)^2 + (y - k)^2 = r^2$

where (h, k) are the coordinates of the center and r is the radius of the circle. For example, the equation:

$(x - 2)^2 + (y - 2)^2 = 25$ or

$(x - 2)^2 + (y - 2)^2 = 5^2$

Represents a circle with the center at (2,2) and radius of 5, as shown below.

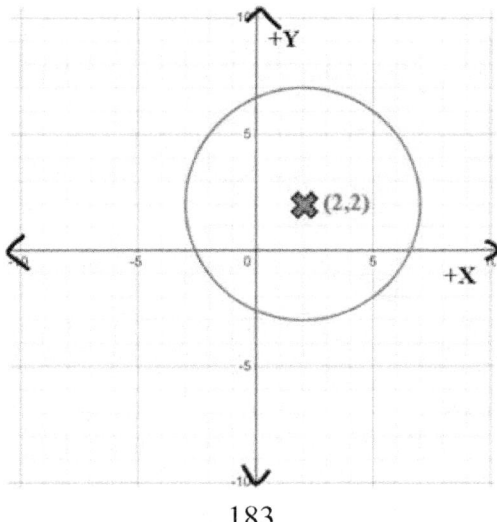

183

Real-World Applications

Circles model circular objects, like Ferris wheels or radar ranges. For example, a radar with center (0, 0) and radius 50 miles has the equation $x^2 + y^2 = 2500$, defining its coverage area.

General form of Circle Equation:

Circle Equation can also be given in the form below:

$x^2 + y^2 + Dx + Ey + F = 0$

This needs to be converted to the standard form below to draw the circle.

$(x - h)^2 + (y - k)^2 = r^2$

We'll show this with a few examples.

Steps to Drawing a Circle Using Standard Form

As mentioned before, the standard form of the equation of a circle is given by:

$(x - h)^2 + (y - k)^2 = r^2$:

Step 1: Identify the center (h, k) and radius r.

Step 2: Plot the center.

184

Step 3: Mark four points on the circle which are the 4 corners of the circle:

$$(h + r, k), (h - r, k), (h, k + r), (h, k - r).$$

Step 4: Draw the circle: Connect points with a smooth curve, using Word's Shapes tool for accuracy.

Now, let's look at an example.

$$(x - 4)^2 + (y - 3)^2 = 36$$

This can be written as:

$$(x - 4)^2 + (y - 3)^2 = 6^2$$

Now, using the standard form below:

$$(x - h)^2 + (y - k)^2 = r^2:$$

Step 1:

We have $h = 4$; $k = 3$; $r = 6$

So, the center of the circle is $(4,3)$ and radius is 6.

Step 2:

Plot the center of the circle $(4,3)$ as shown below:

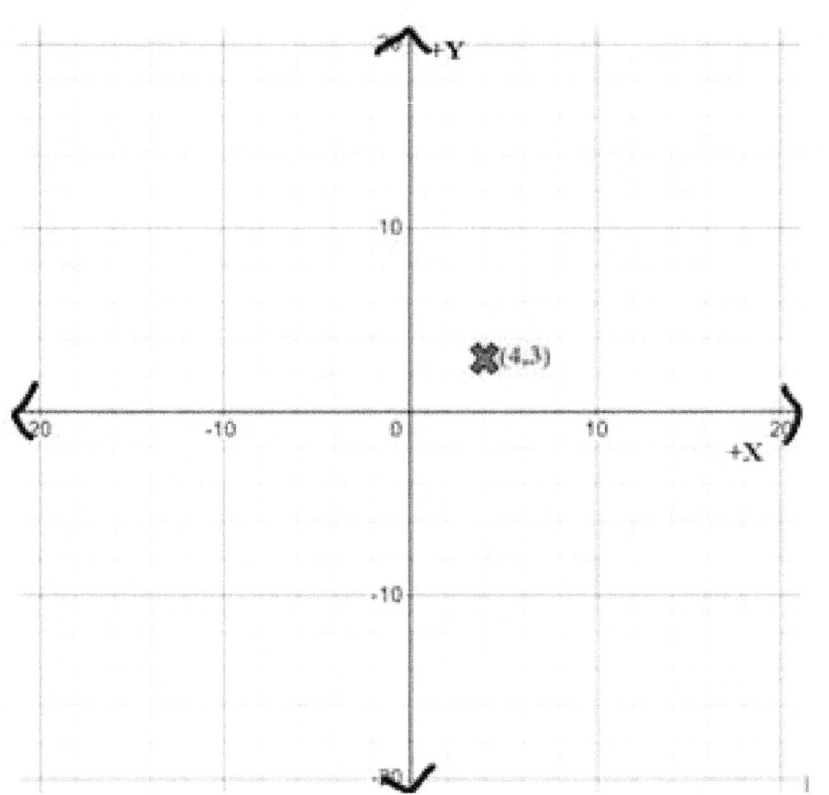

Step 3:

Plot the 4 corners of the circle $(h + r, k)$, $(h - r, k)$, $(h, k + r)$, $(h, k - r)$.

Plugging in $h = 4$, $r = 6$, $k = 3$

The points are $(10, 3)$, $(-2, 3)$, $(4, 9)$, $(4, -3)$.

These points are marked in green below.

186

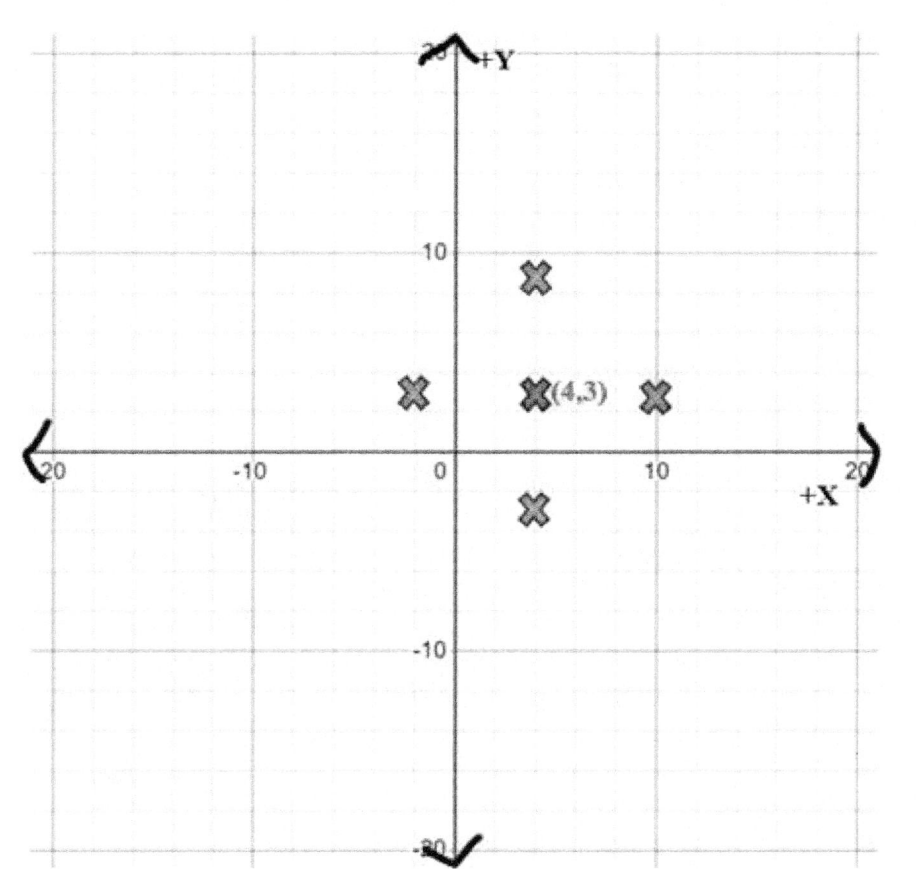

Step 4:

Now join the green points using a compass and create the circle.

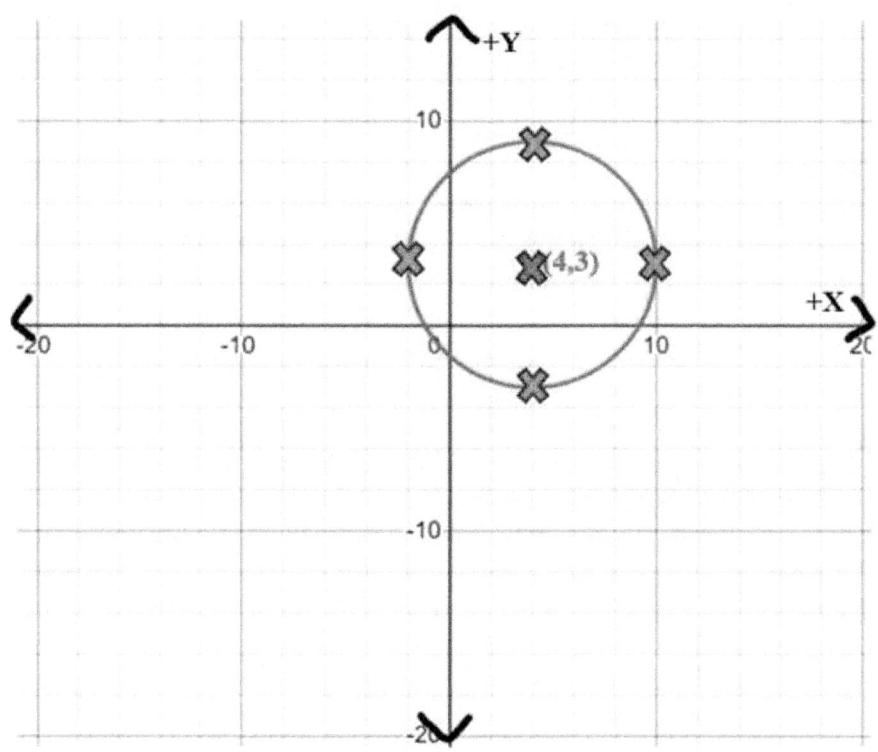

Steps to Drawing a Circle Using The General Form

Now, what if we're given the general form of the equation:

$x^2 + y^2 + Dx + Ey + F = 0$

Rewrite as: $(x^2 + Dx) + (y^2 + Ey) = -F$

Complete the square for x and y, then solve for r^2. This given you the general form equation of the circle. Now, let's have a look at an example.

Example:

Graph $x^2 + y^2 - 6x + 4y - 12 = 0$.

Rewrite grouping the x^2 with x terms; and y^2 with y terms, as shown below:

$(x^2 - 6x) + (y^2 + 4y) = 12$.

Complete the Square:

For x: $x^2 - 6x$ can also be written as

$x^2 - 2.3.x$

Add 9 and subtract 9 to make it a perfect square

$x^2 - 2.3.x + 9 - 9$

$x^2 - 2.3.x + 3^2 - 9$

Now, $x^2 - 2.3.x + 3^2$ and 9 are perfect squares

$(x - 3)^2 - 9$.

Now, do the same for y:

$y^2 + 4y = y^2 + 2.2.y$

$y^2 + 2.2.y + 2^2 - 4$

$(y + 2)^2 - 4$.

So, the equation now becomes

$(x - 3)^2 - 9 + (y + 2)^2 - 4 = 12$.

$(x - 3)^2 + (y + 2)^2 = 25$

Now, the equation is of the general form:

$(x - 3)^2 + (y + 2)^2 = 5^2$

Now, we draw the circle as we did for the standard form.

$(x - h)^2 + (y - k)^2 = r^2$

Step 1:

We have $h = 3$, $k = -2$ and $r = 5$

The circle has a center of $(3, -2)$ and radius 5.

Step 2:

Plot the center of the circle $(3, -2)$.

Step 3:

Plot the 4 corners of the circle $(h + r, k)$, $(h - r, k)$, $(h, k + r)$, $(h, k - r)$.

Plugging in $h = 3$, $r = 5$, $k = -2$

The points are (8, -2), (-2, -2), (3, 3), (3, -7).

Step 4:

Now join the points using a compass and create the circle. All the points and the circle are shown in the graph below.

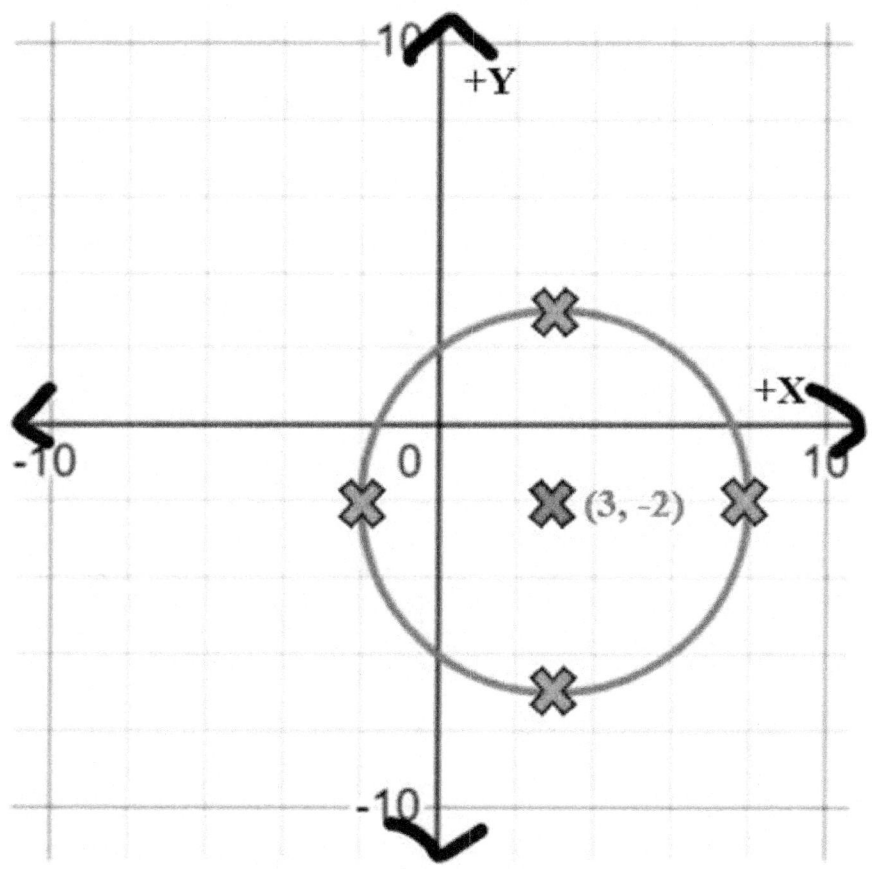

Worksheet 25

Instructions: For each equation, identify the center and radius. If in general form, convert to standard form. Mark the axes and graph the circle on the grid below. Show all work.

1. $(x - 2)^2 + (y + 1)^2 = 9$

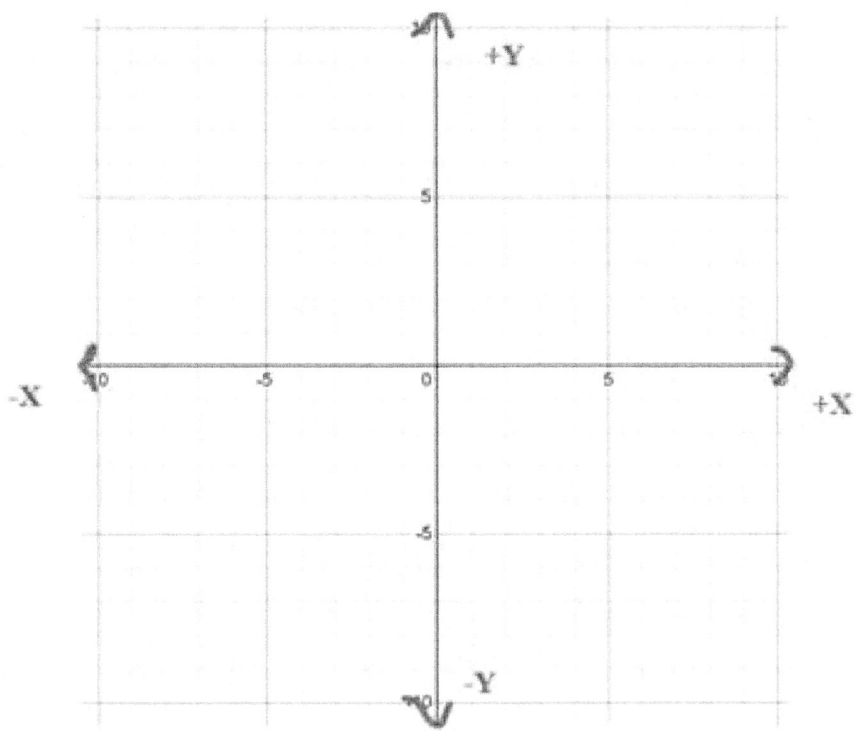

2. $(x - 3)^2 + (y - 4)^2 = 16$

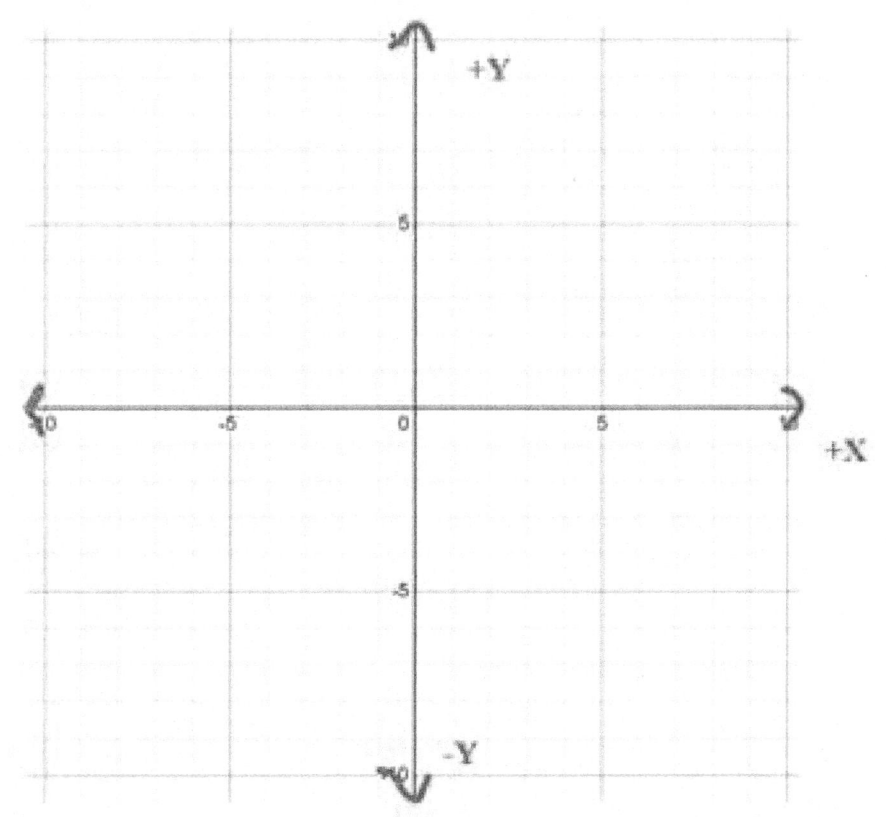

3. $x^2 + y^2 + 4x - 6y - 12 = 0$

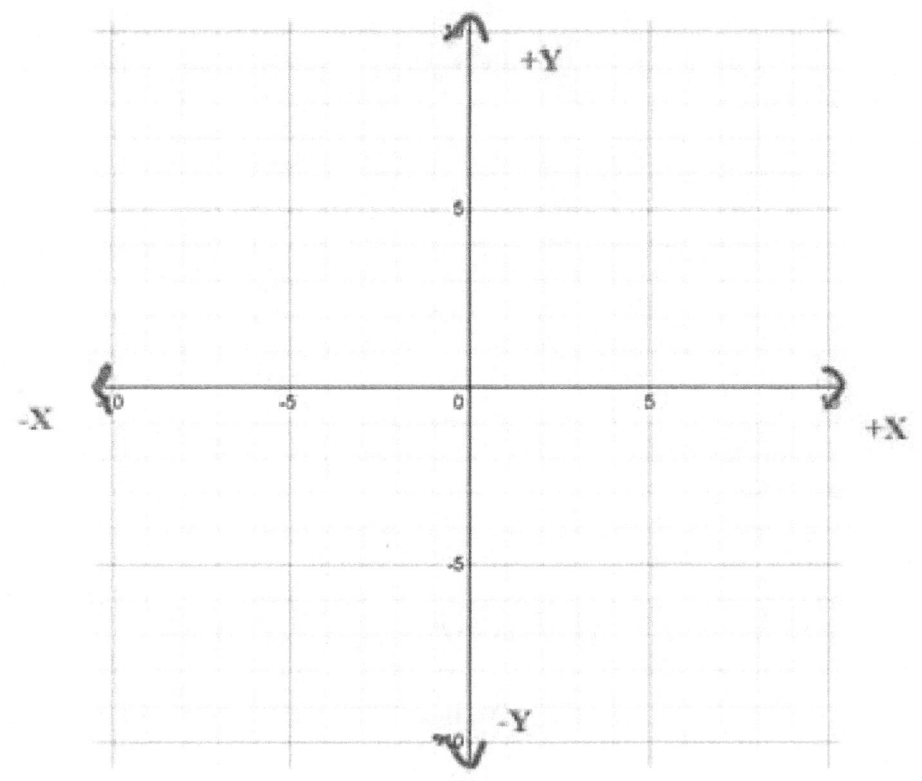

4. $x^2 + y^2 + 6x - 6y - 18 = 0$

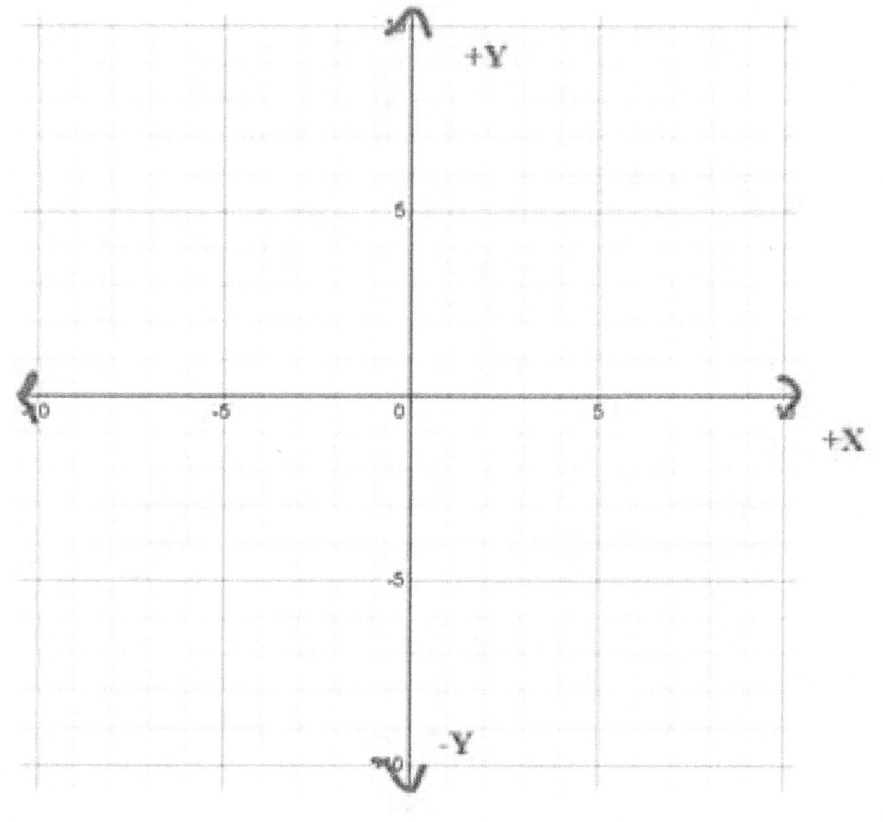

195

5. A circular garden has center at (3, 5) and radius 10 ft. Write its equation in standard form and graph it below.

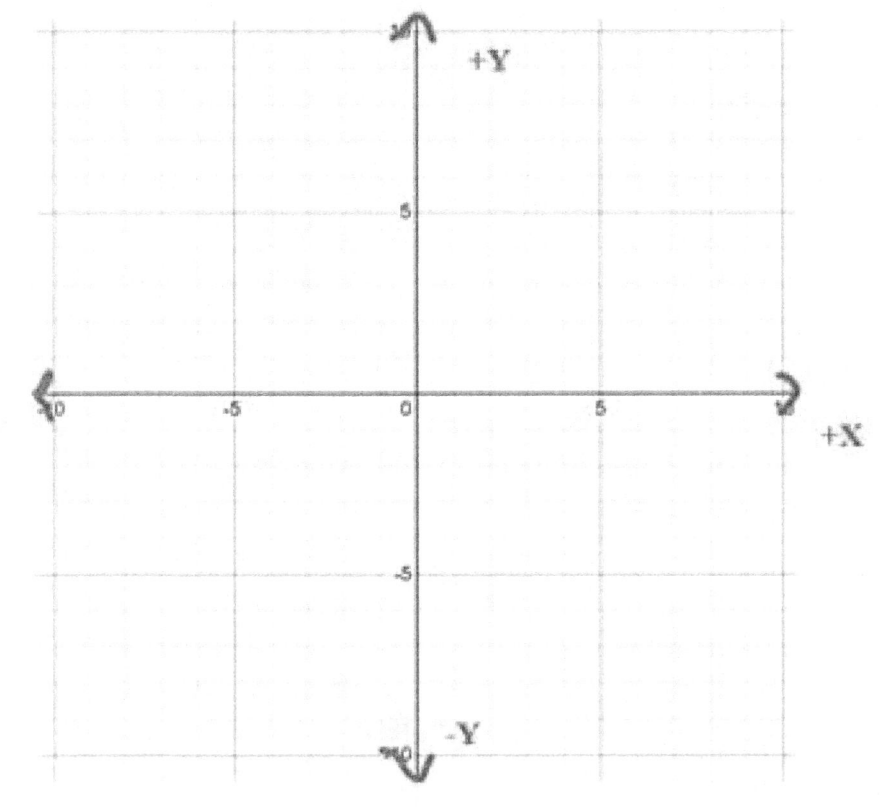

Other Non-Linear Equations

Beyond parabolas and circles, non-linear equations include a variety of curves, each with unique graphs and applications. We'll just touch on them briefly as the details are beyond the scope of this book.

Below are key types:

Ellipses

An ellipse is a stretched circle, with the equation:

$$\frac{(x-h)^2}{a^2} + \frac{(y-k)^2}{b^2} = 1$$

where (h, k) is the center, a is the horizontal radius, and b is the vertical radius. If a > b, the major axis is horizontal; if b > a, it's vertical.

Hyperbolas

A hyperbola consists of two mirror-image curves, with the equation:

$\frac{(x-h)^2}{a^2} + \frac{(y-k)^2}{b^2} = 1$ (horizontal transverse axis)

$\frac{(y-k)^2}{b^2} + \frac{(x-h)^2}{a^2} = 1$ (vertical transverse axis)

The center is (h, k), and asymptotes guide the curve's shape.

Polynomial Functions (Higher Degree)

Higher-degree polynomials, like $y = x^3$, produce complex curves.

Example: $y = x^3 - 4x$ (cubic) has turning points and varying concavity.

Algebraic Inequalities

What are Algebraic Inequalities?

Inequalities are a way to describe relationships between quantities that are not equal. For example, if we have two people named person A and person B with heights of 5.5 feet and 6 ft respectively, we can use an inequality relationship to say that person B is taller than person A.

Height of person B > Height of person A (or)

6 > 5.5

The ">" sign means greater than. And the two quantities we are compared are on both sides of the sign.

An algebraic inequality compares two algebraic expressions that are not of equal value. Unlike Algebraic equations, which have specific solutions, Algebraic inequalities describe a range of solutions. For example, x > 3 means all numbers greater than 3 satisfy the inequality.

Key Features:

- Inequality Symbols:
    - < (Less than)
    - > (Greater than)

- o ≤ (Less than and equal to)

- o ≥. (Greater than and equal to)

- Solution Set: A range of values, often shown on a number line (e.g., x > 3 includes all x from 3 to infinity, excluding 3).

- Linear vs. Non-Linear: Linear Algebraic inequalities (e.g., 2x + 1 < 5) involve variables to the first power; non-linear Algebraic inequalities (e.g., $x^2 > 4$) involve higher powers or products.

Example:

2x - 3 < 7

To find the solution set, we first treat the Algebraic sign ("<") as if it were an "=" sign and solve the equation.

So, we add 3 to both sides.

2x − 3 + 3 < 7 + 3

2x < 10

Dividing both sides by 2, we get:

x < 5

The solution is x < 5, meaning all numbers less than 5 make the inequality true. Here's what it looks like on a number line.

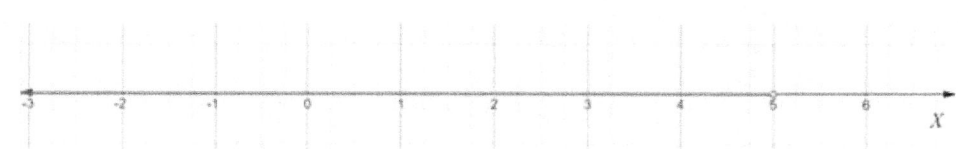

As we can see on above number line, the red line is the solution. It shows that all numbers less than 5 are red. However, we see a circle at 5 that is white, which means that the number 5 is not included in the solution. If the number 5 was included, the circle around 5 would be shaded red as well.

Here's an example that shows it clearly.

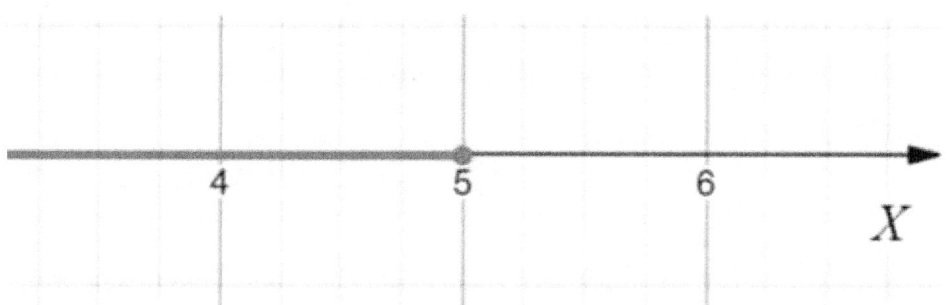

x ≤ 5 (The above number line with shaded red circle)

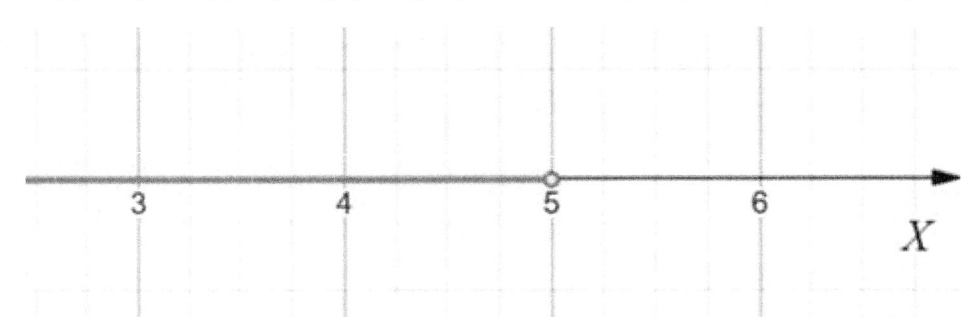

x < 5 (The above number line with hollow red circle)

Real World Examples of Inequalities

Inequalities model situations in the real world where there are constraints that need to be met. The quantities are bounded or compared:

Budgeting: You have $50 to spend on groceries. If g is the cost, $g \leq 50$ ensures you stay within budget.

$g \leq 50$

If you have to buy x fruits and y vegetables and z snacks, the equation becomes:

$x + y + z \leq 50$

Speed Limits: If you have to reach a city y miles away in 5 hours, the car's speed s must satisfy $t \leq 5$ hours to get there on time. So, the inequality becomes:

$t \leq 5$

$\dfrac{y}{x} \leq 5$

Event Planning: If a venue holds up to 100 people, the number of guests g must satisfy $g \leq 100$. If families f have 4 people each, and there are s number of single people, them the equation becomes:

$g \leq 100$

$4f + s \leq 100$

Solving Basic Algebraic Inequalities

As mentioned earlier, solving Algebraic inequalities are very similar to solving Algebraic Equations.

To solve linear inequalities like $ax + b < c$, isolate the variable using the same steps as equations, but with one key rule: **If you multiply or divide both sides by a negative number, reverse the inequality sign.**

So, let's have a look at the steps.

Step 1: Just like we do in a regular equation, combine like terms or clear fractions by cross multiplying.

Step 2: Isolate the Variable: Add/subtract both sides by the same term. Then multiply/divide by the same term if necessary, so we isolate the variable on one side, and constants on the other.

Step 3: This step is the key difference between inequalities and equation. If the variable is negative, we divide both sides by -1. This will flip the +/- sign on both sides of the equation. But we also need to flip the inequality sign since we multiplied/divided by a negative number. So:

< becomes >

> becomes <

≤ becomes ≥

≥ becomes ≤

Step 4: Show the solution on a number line if needed (open circle for < or >, closed circle for ≤ or ≥).

Example:

Solve $3x - 4 \leq 11$.

Step 1,2:

Add 4 to both sides

3x - 4 + 4 ≤ 11 + 4.

3x ≤ 15.

Divide both sides by 3,

x ≤ 5

Step 3: Since the variable is +, we don't need to divide by -1. We just leave the answer as it is.

Step 4: Now we graph it on a number line. The red portion is the solution, and it includes 5, since there is a dark red circle at 5

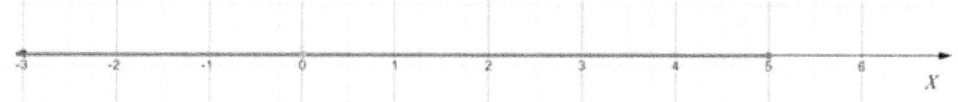

Now, let's look at another example.

Solve:

-2x + 1 > 7.

Step 1,2:

Subtract 1 from both sides:

-2x + 1 - 1 > 7 - 1.

-2x > 6.

Divide both sides by 2.

-x > 3

Step 3: Now we have negative value of the variable. So, we divide both sides by -1. So, we change sides on both sides and also reverse the inequality. '>' becomes '<". We get:

x > -3

Step 4: Now, graph the inequality on a number line.

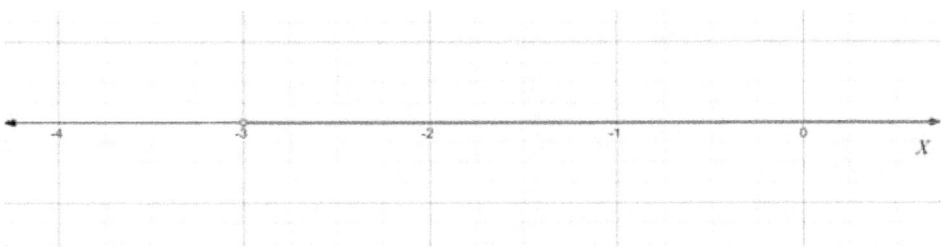

We have an open hollow circle at -3, indicating that -3 is not part of the solution.

Worksheet 26

Solve each inequality and find the range of values that work of x that work for the inequality. Graph the solution on a number line. Show all work.

    1. $4x + 3 < 15$

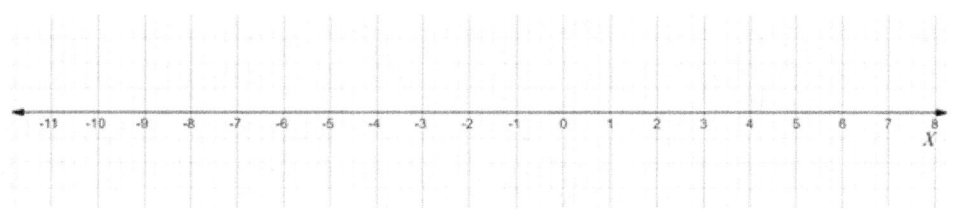

    2. $-3x - 2 \geq 7$

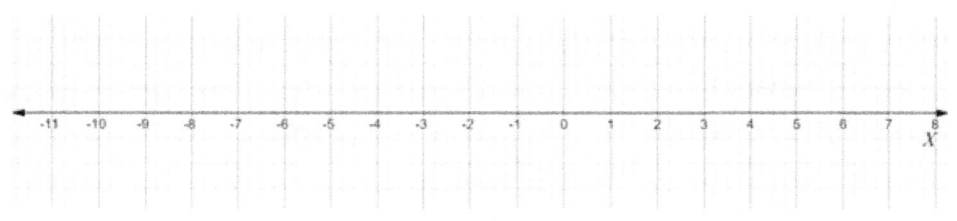

3. $5x - 4 \le 21$

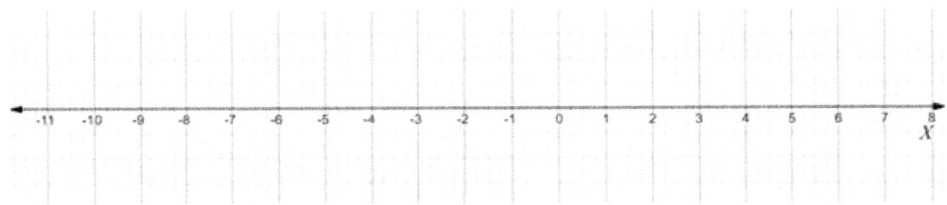

4. $-2x + 9 > 1$

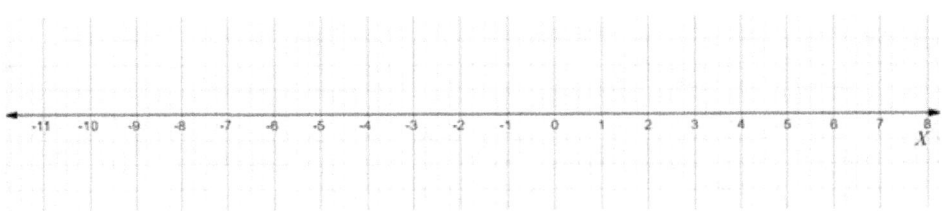

208

5. $7x + 2 \geq -12$

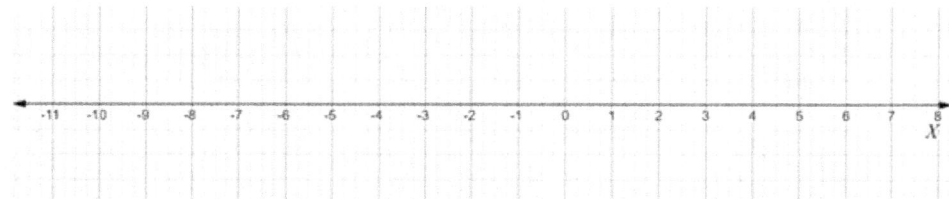

6. $3 - 6x < -15$

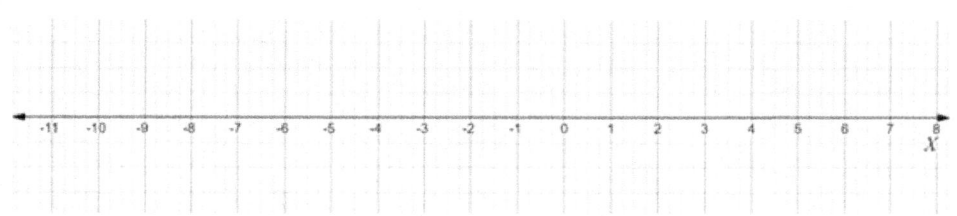

7. $-4x - 5 \leq 11$

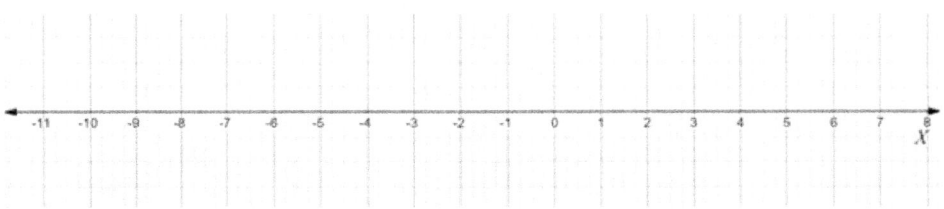

8. $2(x - 3) > 10$

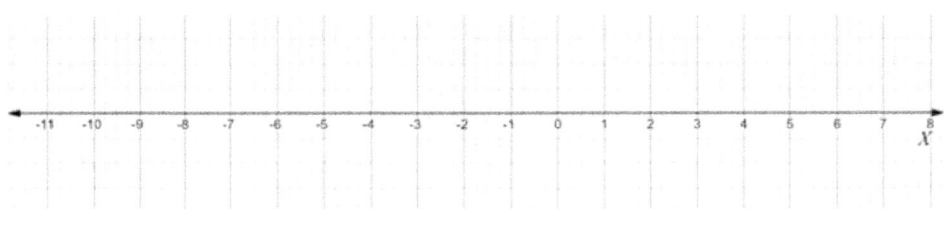

9. A gym membership costs $25 plus $2 per visit. To spend no more than $45, how many visits v can you make? Write and solve an inequality, then graph.

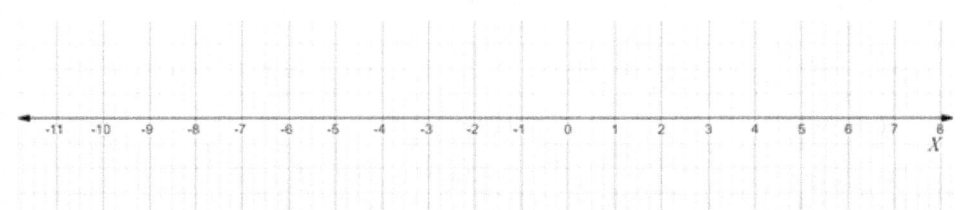

10. Jack has a budget of $10 to get apples and oranges. If Jack buys 2 apples at $0.5 per apple, and oranges cost $0.75 cents each, how many oranges can Jack buy and stay within budget.

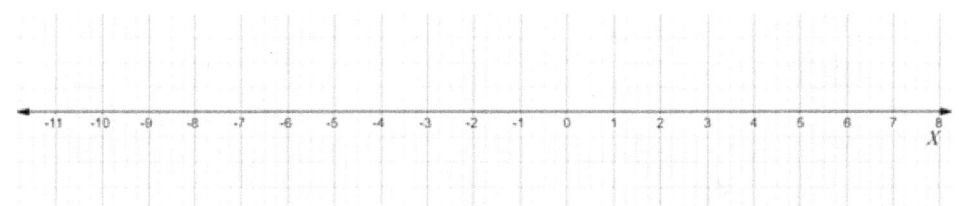

Solving Quadratic Inequalities

A quadratic inequality is an inequality in which one of the terms has a variable of order two. An example would be $ax^2 + bx + c < 0$, which involves a quadratic expression. The solution is a range of x-values where the parabola is below (or above) the x-axis, depending on the inequality.

Below are the steps to solving the expression:

Step 1: Move all terms such that all terms are on side and there is 0 on the other side (e.g., $ax^2 + bx + c < 0$).

Step 2: Find Roots of the quadratic on one side of the equation. Solve $ax^2 + bx + c = 0$ to find x-intercepts (use quadratic formula if needed).

Step 3: Write out the inequality in terms of the roots. For example, if the roots of $ax^2 + bx + c$ are r_1 and r_2, then the inequality becomes:

$$(x - r_1)(x - r_2) < 0$$

Step 4: The roots divide the number line into intervals. Mark the roots r_1 and r_2 on the number line. Test a point in each interval to see if the inequality holds.

The roots r_1 and r_2 are shown on the number line below. Pick a point in each of the boxed regions below and see if they satisfy the inequality.

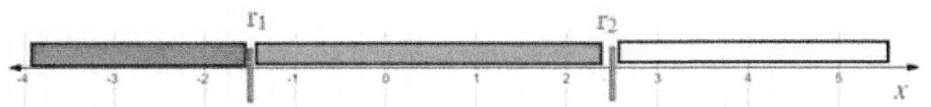

Step 5: Include/Exclude Roots in Solution: For ≤ or ≥, include roots if they satisfy the inequality.

Step 6: Graph: Shade the solution region on a number line.

Example:

Solve:

$x^2 \leq 4x - 3$.

Subtract 4x and add 3 from both sides

$x^2 + 3 - 4x \leq 4x - 3 - 4x + 3$.

$x^2 - 4x + 3 \leq 0$.

Solving the quadratic using the quadratic formula for ax^2+bx+c, we get:

$$x = \frac{-b \pm \sqrt{b^2 - 4ac}}{2a} = \frac{-(-4) \pm \sqrt{-4^2 - 4.1.3}}{2.1} = \frac{4 \pm \sqrt{16 - 12}}{2}$$

$$= \frac{4 \pm 2}{2}$$

x= 3 or x = 1

So, we can rewrite it as:

$(x - 1) (x - 3) \leq 0.$

Roots: $x = 1$, $x = 3$.

So, we mark out the roots 1 and 3 on the number line, and then check the intervals to see if they solve the inequality.

Intervals to test: $(-\infty, 1)$, $(1, 3)$, $(3, \infty)$.

Test Points:

1. The first interval we test is between 1 and $-\infty$ $(-\infty, 1)$. The simplest point is $x = 0$. When we plug $x = 0$ in $(x - 1) (x - 3)$, we get:

   $(0 - 1) (0 - 3) = 3$, which is greater than 0, and the inequality is not sustained in this interval.

2. The second interval we test is between the roots 1 and 3 $(1, 3)$. The simplest point is $x = 2$. When we plug $x = 2$ in $(x - 1) (x - 3)$, we get:

   $(2 - 1) (2 - 3) = -1$, which is less than 0, and the inequality is true in this interval.

3. The third interval we test is between 3 and ∞ $(3, \infty)$. The simplest point is $x = 4$. When we plug $x = 4$ in $(x - 1) (x - 3)$, we get:

(4 - 1) (4 - 3) = 3, which is greater than 0, and the inequality is false in this interval.

So, the solution: 1 < x < 3, Interval: (1, 3).

We represent that on the number line below with closed circles at 1 and 3, line between.

Worksheet 27

Instructions: Solve each quadratic inequality, write the solution in interval notation, and graph it. Show all work.

1. $x^2 - 9 \leq 0$

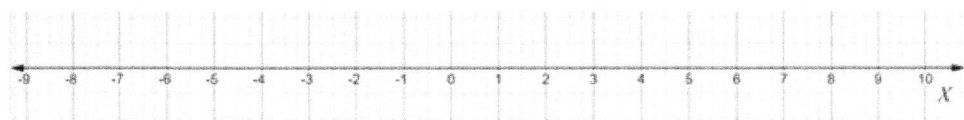

2. $x^2 - 2x - 8 > 0$

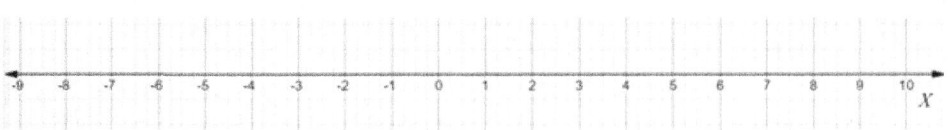

3. $x^2 - 16 \leq 0$

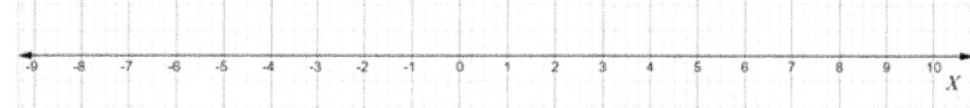

4. $x^2 - 4x - 5 > 0$

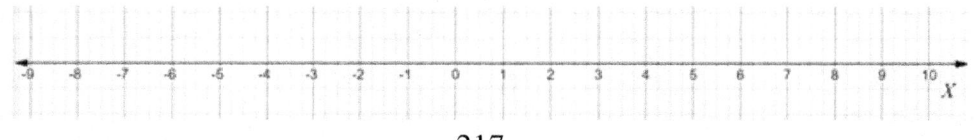

217

5. $2x^2 - 8 \leq 0$

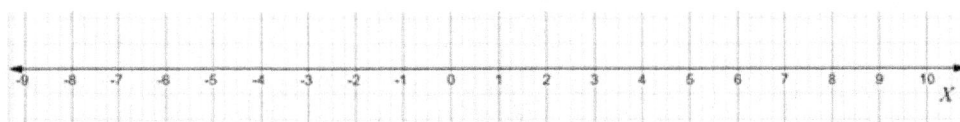

6. $x^2 - 5x + 6 > 0$

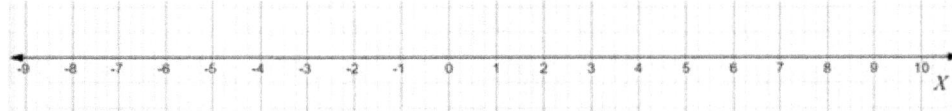

7. $x^2 + 2x - 15 < 0$

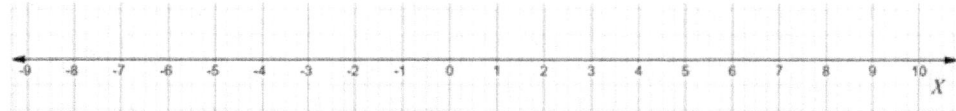

8. $x^2 - x - 6 \geq 0$

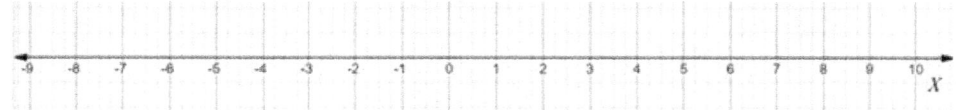

9. A ball is thrown upward from a height of 3 meters with an initial velocity of 14 m/s. Its height after t seconds is given by:

$h(t) = -5t^2 + 14t + 3$

10. A rectangular garden measures 6 meters by 10 meters. You want to build a uniform-width path around it, and you have 36 square meters of material for the path.

Exponent and Logarithmic Equations

What are Algebraic Exponent Equations?

An exponent equation is an equation which has a variable in one of the exponents of the equation. For example, in the equation:

$$a^x = b$$

a is the base and b is the constant. The unknown variable x is in the exponent. These equations model rapid growth or decay, such as compound interest or radioactive decay.

Key Features:

Base: The number raised to a power (e.g., a in a^x).

Exponent: The variable or expression determining the power.

Solution: Often requires isolating the exponential term and may involve logarithms for non-integer exponents.

Solving Algebraic Exponents

To solve $a^x = b$ when the base matches a known value and the exponent is the unknown, there are two ways to solve the equation.

Method A: Match bases (when both sides are powers of the same base).

Method B: Take logarithms (when you cannot easily rewrite both sides with the same base).

Method A: Match bases (Steps for Same Base):

Step 1: Rewrite both sides as powers of the same base.

Step 2: Set the exponents equal.

Step 3: Solve and simplify the resulting equation to get the unknown value.

Now, for example, let's solve:

$3^{(x+1)} = 27.$

Step 1: Rewrite: 27 as 3^3, so $3^{(x+1)} = 3^3$.

Step 2: Equate the exponents, x+ 1 and 3. We get:

$x + 1 = 3.$

Subtracting 1 from both sides,

$x = 2.$

Now, let's do another one. Solve:

$4^x = 64$.

Step 1: Rewrite 16 as 4^3. We get:

Step 2: Equate the exponents:

$x = 3$.

Now, let's do an example where the bases are different. Let's solve:

$4^x = 20$

20 cannot be written as a whole power of 4 so we cannot match bases here.

Take the natural log of both sides

$\ln 4^x = \ln 20$

Based on the power law of logarithms, we get:

$x \ln 4 = \ln 20$

$x = \dfrac{\ln 20}{\ln 4} = 2.17$

Worksheet 28

Solve each exponent equation. Show all work.

1. $2^x = 32$ 2. $3^{2x} = 81$

3. $5^{x-1} = 125$ 4. $4^{x+2} = 16$

5. $10^{3x-4} = 0.01$ 6. $2^x = 15$

7. $7^{2x-3}=50$ 8. $5^{x+1}=42$

9. $9^x=27$ 10. $3^x=20$

11. A population grows by $P = 200 \times 2^{t/5}$, where t is years. When will it reach 800? Solve for t.

Logarithmic Equations

A logarithmic equation is an equation where we take logarithms of the variable, like $\log_a(x) = b$. $\log_a(x)$ is the exponent to which a must be raised to get x. Logarithms are the inverse of exponentials, used in scenarios like earthquake intensity or pH levels.

Key Properties of Logarithms:

$\log_a(b) = c$ means $a^c = b$.

Common Log: $\log_{10}(x)$, written as $\log(x)$.

Natural Log: $\log_e(x)$, written as $\ln(x)$.

Inverse Property: $a^{\log_a(x)} = x$; $\log_a(a^x) = x$.

Product Rule: $\log_a(xy) = \log_a(x) + \log_a(y)$.

Quotient Rule: $\log_a(\frac{x}{y}) = \log_a(x) - \log_a(y)$.

Solving Logarithmic Equations

To solve $\log_a(x) = b$, convert to exponential form or use properties like the product rule:

Steps:

Step 1: Isolate the Logarithm: Get the log term on one side of the equation, and the constant terms on the other.

Step 2: Convert to Exponential: If $\log_a(x) = b$, then $x = a^b$.

Step 3: Solve: Simplify and check for extraneous solutions (x must be positive).

Now, let's look at an example below.

Solve $\log_2(x) = 3$.

Step 1: The logarithm and constant are already isolated.

Step 2: Convert the logarithm to exponential below.

$x = 2^3$.

Step 3: On solving, we get:

$x = 8$.

Now, let's look at another example.

Example:

Solve $\log(x) + \log(x - 3) = 1$.

Step 1: Group the terms together using the log product rule.

Product Rule: $\log(x(x - 3)) = 1$.

Since the log has no base, we assume it is a common log and has a base of 10.

$\log_{10}(x(x - 3)) = 1$

Step 2: Convert to exponential

$x(x - 3) = 10^1 = 10.$

$x^2 - 3x - 10 = 0$

Step 3: Solve the quadratic below:

$x^2 - 3x - 10 = 0$

$x^2 - 5x + 2x - 10 = 0$

$x(x - 5) + 2(x - 5) = 0$

$(x - 5)(x + 2) = 0$

$x = 5$ or $x = -2.$

Worksheet 29

Solve each logarithmic equation. Show all work in the space below.

1. $\log_2 x = 4$

2. $\log_5(x-1) = 2$

3. $\ln x = 3$

4. $\log(x) + \log(x-4) = 1$

5. $\log_3 x + \log_3(x-2) = 2$

6. $\log_7(2x-1) = 1$

7. $\log(x+2) = 0$

8. $\log_2(3x) + \log_2(x-1) = 5$

9. $2\log x = 2$

10. $\log 4x = -1$

Worksheet 30 - Capstone Worksheet

Total: 30 Questions

Time: 45 Minutes

Work out the questions in space provided below:

1. Solve: $x + 7 = 15$

2. Solve: $2x - 5 = 9$

3. Solve: $\frac{x}{3} + 4 = 10$

4. Solve: $3x + 2 = 17$

5. Solve: $x - 8 = -2$

6. Solve: $4x + 3 < 19$

7. Solve: $-2x + 5 \geq 1$

8. Solve: $5x - 7 \leq 13$

9. Solve: $x^2 - 4 = 0$

10. Solve: $x^2 + 5x + 6 = 0$

11. Graph: $y = x^2 - 2x - 3$. Find vertex and intercepts.

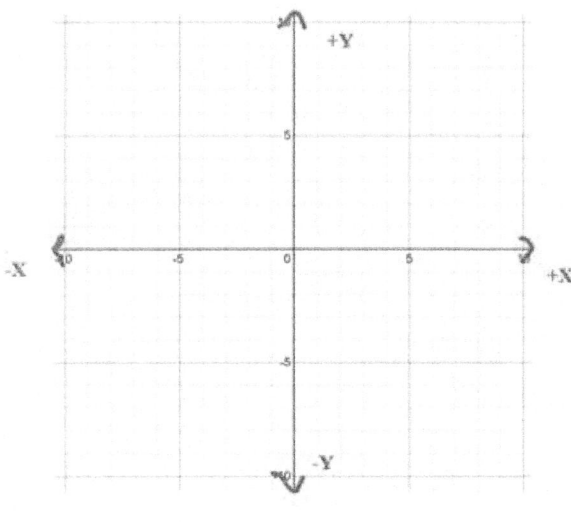

235

12. Graph: $x = y^2 + 2y - 3$. Find vertex and intercepts.

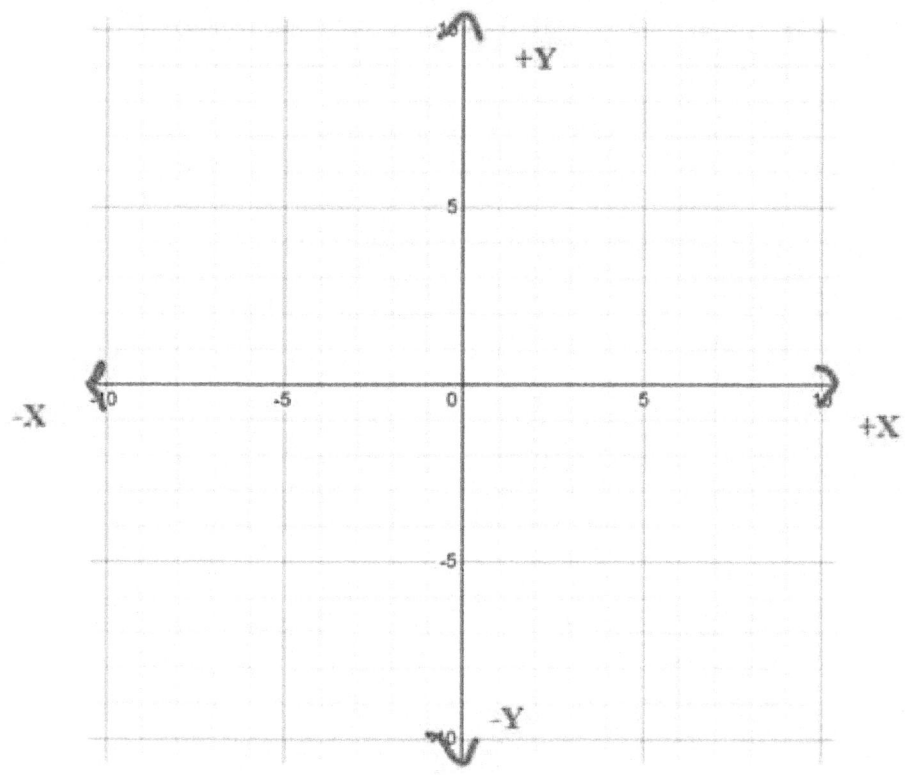

13.   Solve: $(x - 2)^2 + (y + 1)^2 = 9$. Identify center and radius.

Convert: $x^2 + y^2 - 4x + 6y - 12 = 0$ to standard circle form.

14.   Solve: $2^x = 16$

15.   Solve: $3^{x+1} = 27$

16. Solve: $\log_2 x = 5$

17. Solve: $\log_3(x - 2) = 4$

18. Solve: $\ln x = 2$

19. Solve: $\log(x) + \log(x - 3) = 1$.

20. A car travels 120 km in x hours at 60 km/h. Solve for x

21. If $3x + 4$ shirts cost $50, find x (cost per shirt).

22. A rectangle's length is $x + 2$, width is x. Area is 24. Solve for x.

23. If a budget consists of buying x items and and y items, the total budget is given by: $2x + 3y \leq 30$, $x, y \geq 0$. Maximize the value of x if $y = 5$.

24. The height of an item increases with time by the equation:

Height: $h = -5t^2 + 20t + 2$. Find max height.

25. Solve: $x^2 - 6x + 8 \leq 0$

26. Solve: $x^2 + 2x - 3 > 0$

27. Population: $P = 100 \times 2^t$. When is $P = 800$?

28. A path costs \$10/m², total \$50. Area $x^2 \leq 5$. Solve for x.

29. Verify: $2x - 3 = 7$, proposed $x = 5$.

Answers

Worksheet 1:

(1) Yes (2) No (3) Yes (4) No (5) Yes

Worksheet 2:

(1) x -12 = 5; (2) 3y + 7 = 22; (3) 2x + 9 = 21; (4) x/2 + 5 = 14;
(5) x/4 − 3 = 8; (6) 3s + 5 = 50; (7) 2s = 4; (8) 50t = 200; (9) 20w = 180; (10) 12t = $84

Worksheet 3:

(1)x=17 (2)x=5 (3)x=18 (4)x=44 (5)x=15 (6)x=6 (7)s=2 (8)t=4 (9)w = 9 (10)t=7

Worksheet 4:

(1) True (2) False (3) True (4) True (5) True (6) False (7) True (8) True (9) True (10) False

Worksheet 5:

(1)7 (2) a (3) a,b (4) 0 (5) y+a (6) 6x + 30 (7) 0 (8) 1/y (9) 2y + 12 (10)27

Worksheet 6:

(1)8 (2)25 (3)10000 (4)3 (5)1 (6)3^6 (7)4^7 (8)6^4 (9)9^3 (10)2^6 (11)2^6 (12)5^8 (13)3^6 (14)7^6 (15)1/64 (16)1/64 (17)1/10 (18)1/9 (19)1 (20) 5

Worksheet 7:

(1) Single Variable Equation, Linear (2) Quadratic Equation
(3) Multi-Variable Equation, Linear (4) Rational Equation (5) Higher Order
Equation (Cubic) (6) Exponential Equation (7) Logarithmic Equation (8)
Single Variable Equation, Linear (9) Higher Order Equation (Quartic) (10)
Multi-Variable Equation, Linear (11) Quadratic Equation (12) Exponential
Equation (13)Logarithmic Equation (14) Rational Equation (15) Higher Order
Equation (Quintic)

Worksheet 8:

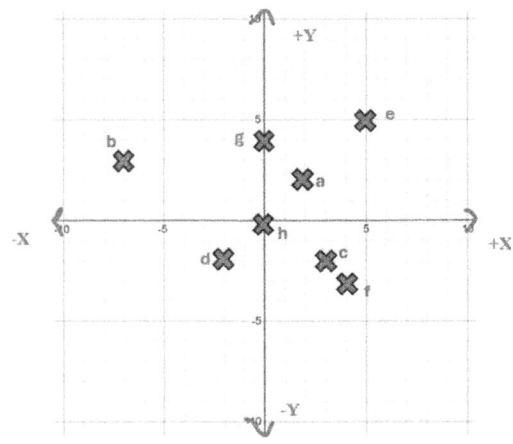

Worksheet 9:

1. a. No b. Yes c. Yes d. No e. Yes
2. a. $y = 1$, b. $y = 7$, c. $y = 0$, d. $x = 1/3$, e. $x = -1/3$

Worksheet 10:

1. (a) 3 (b) 2 (c) -2/3 (d) 7 (e) -2

3.   (a) 1, (b) 1.5, (c) 1, (d) 0.8, (e) 4

Worksheet 11:

a)   4, b) 7, c) -5, d) 0 e) 0 f) 2, g) -3, h) 6, g) -3, h) 6, i) 9, j) -1

Worksheet 12:

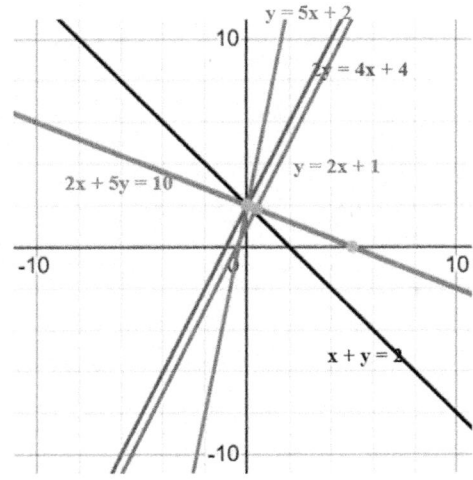

Worksheet 13:

1. (a)2 (b)2; 2.(a)1 (b) 0; (3) c = 2; (4) y = 3x + 2; (5) c = -1 (6) y = -2x – 1

(7) m = -2; (8) y = -2x + 5; (9)y = -2x – 6; (10) y = (7/3)x + 2

Worksheet 14:

1. 29/5, 2. -22, 3. -11/2, 4. -8, 5. 15

Worksheet 15:

(1) a. 6, b. 20, c. 2, d. 21, e. 4, f. 10

245

(2) a. 5/6, b. (5x+2)/10, c. (3x + 10)/6, d. 9x/14, e. 23x/20,

f. (3x+4)/2x

(3) a. -3/5, b. -7/6, c. -15, d. 5/24, e. 2, f. 11/2 g. -5/11, h. 1, g. 3, h. 24

Worksheet 16:

(1) x = 4, y = 3 (2) x = 3, y = 1 (3) x = 1, y = 1 (4) x = 2, y = 0

(5) 3 pens and 4 notebooks

Worksheet 17:

(1) x = 2.5, y = 1 (2) x = 2, y = 1 (3) x = 2, y = 1 (4) x = 5/3; y = 1

(5) 30 small coffees; 20 large coffees

Worksheet 18:

(1) x = 3, y = 1 (2) x = 3, y = 0 (3) x = 2, y = 1 (4) x = 5/4; y = 3/2 (5) 5
muffins; 5 cookies

Worksheet 19:

(1) (a) 4 (2x + 3) (b) 5y (3y - 5) (c) $3a^2$ (2a + 3) (d)7mn (2m + 3n)
(e) ab (b + a + c)

(2) 12 fruits a basket

(3) 12xy $(2x^2y - 3x + y)$

(4) (a) (x + y) (2 + a) (b) (m − n) (3 + k) (c) (x + y) (2 + x)
(d) (2a − b) (3 + b) (e) (x + 2) (x − y)

(5) $2(x^2 + y) (x + 1)$

(6) (2x + 1) and (x + y)

Worksheet 20:

1. $(x + 3)(x + 4)$
2. $(x - 3)(x - 5)$
3. $(3x + 2)(x + 3)$
4. $(2x + 1)(x + 3)$
5. $(x + 6)(x - 5)$
6. $(x + 2)(x + 3)$
7. $(x - 4)(x - 2)$
8. $(2x + 1)(x - 3)$
9. $(3x - 4)(x - 2)$
10. $(x + 5)(x - 2)$

Worksheet 21:

1) $(x + 3)^2$
2) $(2x + 3)^2$
3) $(x - 5)^2$
4) $(3x + 1)^2$
5) $(4x - 3)^2$
6) $(x + 2)(x - 2)$
7) $(3x + 5)(3x - 5)$
8) $(2x + 7)(2x - 7)$
9) $(x + 10)(x - 10)$
10) $(5x + 4)(5x - 4)$
11) $(x - 4)(x^2 + 4x + 16)$
12) $27(x+2)(x^2 - 2x + 4)$

Worksheet 22:

1. $x = -3$ and $x = -5$
2. $x = 3$ and $x = 4$
3. $x = 5$ (Repeated)
4. $x = -1$ and $x = -3/2$
5. $x = 1$ and $x = 2$

Worksheet 23:

1. $x = 3$ and $x = -1$
2. $x = -1$ (repeated)
3. $x = -0.38$ or $x = -2.62$
4. $x = 3.58$ or $x = 0.42$
5. $x = 1$
6. $x = 0.62$ and $x = -1.62$
7. $x = 1$ and $x = 2/5$
8. $x = 2$ (Repeated)
9. $x = 3$ and $x = -0.5$
10. $x = -0.44$ and $x = -4.57$

Worksheet 24:

1. $y = x^2 + 2x - 3$

   Vertex: (-1, -4); y-intercept: (0, -3), x-intercepts: (1, 0), (-3, 0).

2. $y = -2x^2 + 8x - 6$

   Vertex: (2, 2), y-intercept: (0, -6), x-intercepts: (1, 0), (3, 0).

3. $x = 2y^2$

   Vertex: (0, 0); y-intercept: (0, 0), x-intercepts: (0, 0).

4. $x = y^2 + 4y + 3$

   Vertex: (-1, -2); y-intercept: (0, -3),(0,-1); x-intercepts: (3, 0).

5. Rocket Vertex: (2, 37), max height 37 ft at t = 2 sec. Hits ground at t ≈ 4.15 sec.

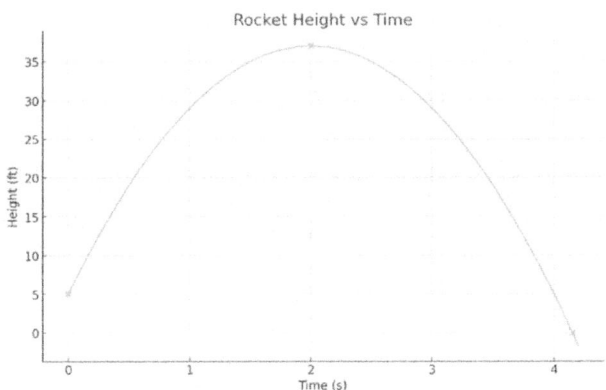

Worksheet 25:

1. Center: (2, -1), Radius: 3

2. Center: (3,4), Radius: 4.

3. Center: (-2, 3), Radius: 5.

4. Center: (-3,3), Radius: 6.

5. Garden: Equation: $(x - 3)^2 + (y - 5)^2 = 100$.

Worksheet 26:

1. $x < 3$.

2. $x \leq -3$.

3. $x \leq 5$

4. $x < 4$

5. $x \geq -2$

6. $x > 3$

7. $x \geq -4$

8. $x > 8$

9. The number of visits $v \leq 10$

10. The number of oranges $o \leq 12$

Worksheet 27:

1. $[-3, 3]$ or $-3 < x < 3$

2. $(-\infty, -2) \cup (4, \infty)$ or $x < -2 \cup x > 4$.

3. $[-4, 4]$ or $-4 < x < 4$

4. $(-\infty, -1) \cup (5, \infty)$ or $x < -1 \cup x > 5$.

5. $[-2, 2]$ or $-2 < x < 2$

6. $(-\infty, 2) \cup (3, \infty)$ or $x < 2 \cup x > 3$.

7. $[-5, 3]$ or $-5 < x < 3$

8. $(-\infty, -2) \cup (3, \infty)$ or $x < -2 \cup x > 3$.

9. Between time of 0.65 seconds and 3 seconds

10. Maximum width of 1 metre.

Worksheet 28:

1. x = 5
2. x = 2
3. x = 4
4. x = 0
5. x = 2/3
6. x = 3.907
7. x ≈2.666
8. x ≈1.583
9. x = 1.5
10. x ≈ 2.727
11. 10 years

Worksheet 29

1. 16
2. 26
3. e3
4. 5.74
5. 4.16
6. 4
7. -1
8. 3.804
9. 10

10. 0.025

Worksheet 30:

1. $x = 8$

2. $x = 7$

3. $x = 18$

4. $x = 5$

5. $x = 6$

6. $x < 4$

7. $x \leq 2$

8. $x \leq 4$

9. $x = +2, -2$

10. $x = -2, -3$

11. Vertex: (1, -4), y-intercept: (0, -3), x-intercepts: (3, 0), (-1, 0)

12. Vertex: (-4, -1), x-intercept: (0, -3), y-intercepts: (0, -3), (0,1)

13. Center: (2, -1), Radius: 3

14. $(x - 2)^2 + (y + 3)^2 = 25$

15. $x = 4$

16. $x = 2$

17. $x = 32$

18. $x = 83$

19. $x = e^2 \approx 7.39$

20. $x = 5$

21. $x = 2$

22. $x \approx \$15.33$

23. $x = 4$

24. $x = 7.5$

25. Max height: 22 at $t = 2$

26. $2 \leq x \leq 4$

27. $x < -3$ or $x > 1$

28. $t = 3$

29. $x \leq 5 \approx 2.23$

30. Verified: $2(5) - 3 = 7$, correct

Bonus #3 – We'll Solve A Problem for Free

Email us at **abiprod.pty.ltd@gmail.com** with any Math Problem you need solved (related to the content in this book). We'll solve one problem for you for free.

Printed in Dunstable, United Kingdom

67362104R00147